Ad Hoc Mobile Wireless Networks

Principles, Protocols, and Applications

Second Edition

OTHER TELECOMMUNICATIONS BOOKS FROM AUERBACH

Ad Hoc Mobile Wireless Networks

Principles, Protocols, and Applications

Second Edition

Subir Kumar Sarkar
T.G. Basavaraju
C. Puttamadappa

CRC Press
Taylor & Francis Group
Boca Raton   London   New York

CRC Press is an imprint of the
Taylor & Francis Group, an **informa** business

CRC Press
Taylor & Francis Group
6000 Broken Sound Parkway NW, Suite 300
Boca Raton, FL 33487-2742

First issued in paperback 2016

© 2013 by Taylor & Francis Group, LLC
CRC Press is an imprint of Taylor & Francis Group, an Informa business

No claim to original U.S. Government works

Version Date: 20120725

ISBN 13: 978-1-138-19881-4 (pbk)
ISBN 13:978-1-4665-1446-1(hbk)

Library of Congress Cataloging-in-Publication Data

Sarkar, Subir Kumar.
    Ad hoc mobile wireless networks : principles, protocols, and applications / Subir Kumar Sarkar, T.G. Basavaraju, C. Puttamadappa. -- Second edition.
    pages cm
    Includes bibliographical references and index.
    ISBN 978-1-4665-1446-1 (hardback)
    1. Wireless communication systems--Quality control. 2. Internetworking (Telecommunication) 3. Ad hoc networks (Computer networks) I. Basavaraju, T. G. II. Puttamadappa, C. III. Title.

    TK5103.2.S27 2013
    004.6--dc23   2012020880

Visit the Taylor & Francis Web site at
http://www.taylorandfrancis.com

and the CRC Press Web site at
http://www.crcpress.com

Contents

1

INTRODUCTION

1.1 Fundamentals of Wireless Networks

Communication between various devices makes it possible to provide unique and innovative services. Although this interdevice communication is a very powerful mechanism, it is also a complex and clumsy mechanism, leading to a lot of complexity in present day systems. This not only makes networking difficult but also limits its flexibility. Many standards exist today for connecting various devices. At the same time, every device has to support more than one standard to make it interoperable between different devices. Take the example of setting up a network in offices. Right now, entire office buildings have to make provisions for lengths of cable that stretch kilometers through conduits in the walls, floors, and ceilings to workers' desks.

In the last few years, many wireless connectivity standards and technologies have emerged. These technologies enable users to connect a wide range of computing and telecommunications devices easily and simply, without the need to buy, carry, or connect cables. These technologies deliver opportunities for rapid ad hoc connections, and the possibility of automatic, unconscious connections between devices. They will virtually eliminate the need to purchase additional or proprietary cabling to connect individual devices, thus creating the possibility of using mobile data in a variety of applications. Wired local area networks (LANs) have been very successful in the last few years and now, with the help of these wireless connectivity technologies, wireless LANs (WLANs) have started emerging as much more powerful and flexible alternatives to the wired LANs. Until a year ago, the speed of the WLAN was limited to 2 megabits per second (Mbps), but with the introduction of these new standards, we are seeing WLANs that can support up to 11 Mbps in the industrial, scientific, and medical (ISM) band.

There are many such technologies and standards, and notable among them are Bluetooth, Infrared Data Association (IrDA), HomeRF, and Institute of Electrical and Electronic Engineers (IEEE) 802.11 standards. These technologies compete on certain fronts and are complementary in other areas. So, given the fact that so many technologies exist, which technology is the best and which solution should one select for a specific application? To be able to understand this, we need to look at the strengths and weaknesses and also the application domains of each of these standards and technologies.

The premise behind all these standards is to use some kind of underlying radio technology to enable wireless transmission of data, and to provide support for formation of networks and managing various devices by means of high-level software. Bluetooth, though quite new, has emerged as the forerunner in this so-called "battle between competing technologies" due to the kind of support it is getting from all sections of the industry. However, it must be kept in mind that the viability of a technology depends on the application context.

1.1.1 Bluetooth

Bluetooth is a high-speed, low-power microwave wireless link technology designed to connect phones, laptops, personal digital assistants (PDAs), and other portable equipment with little or no work by the user. Unlike infrared, Bluetooth does not require line-of-sight positioning of connected units. The technology uses modifications of existing WLAN techniques but is most notable for its small size and low cost. Whenever any Bluetooth-enabled devices come within range of each other, they instantly transfer address information and establish small networks between each other, without the user being involved.

Bluetooth is an open wireless technology standard for exchanging data over short distances from fixed and mobile devices, creating personal area networks (PANs) with high levels of security. It was created by telecom vendor Ericsson in 1994. It was originally thought of as a wireless alternative to RS-232 data cables. It can connect several devices and overcomes the problems of synchronization.

At any given time, data can be transferred between the master and one other device. The master chooses which slave device to address.

It switches rapidly from one device to another in a round-robin fashion. Since it is the master that chooses which slave to address, a slave is supposed to listen in each receive slot. Being a master is a lighter burden than being a slave. Being a master of seven slaves is possible; being a slave of more than one master is difficult.

Features of Bluetooth technology include the following:

- Operates in the 2.56 GHZ ISM band, which is globally available (no license required)
- Uses frequency hop spread spectrum (FHSS)
- Can support up to eight devices in a small network known as a "piconet"
- Omnidirectional, non-line-of-sight transmission through walls
- 10–100 m range
- Low cost
- 1 mW power
- Extended range with external power amplifier (100 m)

1.1.2 IrDA

IrDA is an international organization that creates and promotes interoperable, low-cost infrared data interconnection standards. IrDA has a set of protocols covering all layers of data transfer and, in addition, has some network management and interoperability designs. IrDA protocols have IrDA DATA as the vehicle for data delivery and IrDA CONTROL for sending the control information. In general, IrDA is used to provide wireless connectivity technologies for devices that would normally use cables for connectivity. IrDA is a point-to-point, narrow angle (30° cone), ad hoc data transmission standard designed to operate over a distance of 0–1 m and at speeds of 9600 bits per second (bps) to 16 Mbps. Adapters now include the traditional upgrades to serial and parallel ports.

Features of IrDA are as follows:

- Range: from contact to at least 1 m and can be extended to 2 m; a low-power version relaxes the range objective for operation from contact through at least 20 centimeters (cm)

between low-power devices and 30 cm between low-power
and standard-power devices. This implementation affords 10
times less power consumption.
- Bidirectional communication is the basis of all specifications.
- Data transmission from 9600 bps with primary speed or cost
steps of 115 kilobits per second (kbps) and maximum speed
of up to 4 Mbps.
- Data packets are protected using a cyclic redundancy check
(CRC) (CRC-16 for speeds up to 1.152 Mbps, and CRC-32
at 4 Mbps).

1.1.2.1 Comparison of Bluetooth and IrDA   Bluetooth and IrDA are
both critical to the marketplace. Each technology has advantages and
drawbacks and neither can meet all users' needs. Bluetooth's ability to penetrate solid objects and its capability for maximum mobility within the piconet allow for data exchange applications that are
very difficult or impossible with IrDA. For example, with Bluetooth,
a person could synchronize his or her phone with a personal computer (PC) without taking the phone out of a pocket or purse; this is
not possible with IrDA. The omnidirectional capability of Bluetooth
allows synchronization to start when the phone is brought into range
of the PC.

On the other hand, in applications involving one-to-one data
exchange, IrDA is at an advantage. Consider an application where
there are many people sitting across a table in a meeting. Electronic
cards can be exchanged between any two people by pointing the IrDA
devices toward each other (because of the directional nature). In contrast, because Bluetooth is omnidirectional in nature, the Bluetooth
device will detect all similar devices in the room and the user would
have to select the intended person from, say, a list provided by the
Bluetooth device. On the security front, Bluetooth provides security mechanisms that are not present in IrDA. However, the narrow
beam (in the case of IrDA) provides a low level of security. IrDA beats
Bluetooth on the cost front. The Bluetooth standard defines layers 1
and 2 of the open system interconnection (OSI) model. The application framework of Bluetooth is aimed to achieve interoperability with
IrDA and wireless access protocol (WAP). In addition, a host of other
applications will be able to use the Bluetooth technology and protocols.

1.1.3 HomeRF

HomeRF is a subset of the International Telecommunication Union (ITU) and primarily works on the development of a standard for inexpensive radio frequency (RF) voice and data communication. The HomeRF Working Group has also developed the shared wireless access protocol (SWAP). SWAP is an industry specification that permits PCs, peripherals, cordless telephones, and other devices to communicate voice and data without the use of cables. SWAP is similar to the carrier sense multiple access with collision avoidance (CSMA/CA) protocol of IEEE 802.11, but with an extension to voice traffic.

The SWAP system can operate either as an ad hoc network or as an infrastructure network under the control of a connection point. In an ad hoc network, all stations are peers and control is distributed between the stations and supports only data. In an infrastructure network, a connection point is required so as to coordinate the system and it provides the gateway to the public switched telephone network (PSTN). Walls and floors do not cause any problem in its functionality and some security is also provided through the use of unique network IDs. It is robust and reliable, and it minimizes the impact of radio interference.

Features of HomeRF are as follows:

- Operates in the 2.45 GHz range of the unlicensed ISM band
- Range: up to 150 feet
- Employs frequency hopping at 50 hops per second
- Supports both a time division multiple access (TDMA) service to provide delivery of interactive voice and CSMA/CA service for delivery of high-speed data packets
- Capable of supporting up to 127 nodes
- Transmission power: 100 mW
- Data rate: 1 Mbps using 2 frequency-shift keying (FSK) modulation and 2 Mbps using 4 FSK modulation
- Voice connections: up to six full duplex conversations
- Data security: blowfish encryption algorithm (over 1 trillion codes)
- Data compression: Lempel-Ziv Ross Williams 3 (LZRW3)-A algorithm

1.1.3.1 Comparison of Bluetooth with Shared Wireless Access Protocol (SWAP) Currently, SWAP has a larger installed base compared to Bluetooth, but it is believed that Bluetooth is eventually going to prevail. Bluetooth is a technology to connect devices without cables. The intended use is to provide short-range connections between mobile devices and to the Internet via bridging devices to different networks (wired and wireless) that provide Internet capability. HomeRF SWAP is a wireless technology optimized for the home environment. Its primary use is to provide data networking and dial tones between devices such as PCs, cordless phones, Web tablets, and a broadband cable or digital subscriber line (DSL) modem. Both technologies share the same frequency spectrum but do not interfere with each other when operating in the same space. As far as comparison with IrDA is concerned, SWAP is closer to Bluetooth in its scope and domain, so the comparison between Bluetooth and IrDA holds good to a large extent between these two also. Comparisons of these technologies are given in Table 1.1.

Wireless networks use finite resources, and a given geographical area with many wireless networks will degrade in performance as more users come on. For example, a building with 20 competing networks can cause interference and slow performance for all users. Wireless networks are flexible and can be deployed quickly using inexpensive radio equipment and antennas. The flexibility of being able to deploy a network rapidly means that many networks operating in the same area can "peer" or aggregate themselves into a larger network with more capacity to be used by users.

Table 1.1 Comparison of Various Wireless Technologies

	PEAK DATA RATE	RANGE	RELATIVE COST	VOICE NETWORK SUPPORT	DATA NETWORK SUPPORT
IEEE 802.11	2 Mbps	50 m	Medium	Via Internet protocol (IP)	Transmission control protocol (TCP)/IP
IrDA	16 Mbps	<2 m	Low	Via IP	Via point-to-point protocol (PPP)
Bluetooth	1 Mbps	<10 m	Medium	Via IP and cellular	Via PPP
HomeRF	1.6 Mbps	50 m	Medium	Via IP and PSTN	TCP/IP

Wireless networks act in a similar manner to people discussing something in a public area. The discussion can be "heard" by others in the area with appropriate equipment. Security issues are thus pushed to the users, forcing the use of encryption and "safe computing" practices that are generally avoided by the public at large today. Wireless network speeds do not (yet) fare well against the gigabit speeds achieved by wired networks such as gigabit Ethernet or fiber. However, wireless network technology is rapidly maturing, and new, open standards are emerging that will provide speeds comparable to those of fiber and other infrastructures. Wireless network technologies based on IEEE 802.11 and 802.16 standards (wireless fidelity [WiFi] and worldwide interoperability for microwave access [WiMax]) are not restricted to any one vendor and can be deployed by anyone with a basic understanding of the technology. Wireless networks are ideal for connecting many people without the expenses of deploying cable and human resources. Wireless networks provide mobility and access to information based on physical proximity.

A typical wireless network consists of (1) an access point and (2) client wireless radios used by each subscriber. The access is a "central hub" device that provides service to 1–100 subscribers. Multiple access points may be required in larger geographic areas or to serve large groups of users. An access point can be connected to other access points or connected directly to the network that provides the connection to the Internet in one's community. The access point is typically placed in a central location within view of a group of subscribers and within view of other access points or with a network link to a point of presence (POP).

The access point manages the flow of information between subscribers and to other elements in the network. It broadcasts a network service set ID (SSID), or network name, and handles limited security functions. When a subscriber links to the community wireless network, his or her subscriber radio is configured to use the access point's SSID and relevant security parameters. The subscriber radio then establishes a connection to the wireless network, and a data connection is created.

A computer system is connected to a wireless device using an Ethernet cable. Information sent from the computer (or other computers on the same Ethernet network) is delivered to the wireless device:

- A transmitter sends radio signals with information to an antenna.
- The antenna takes the radio signals and directs them into the air and directs the radio signal toward a specific physical location.
- A receiver hears the radio signals by way of its own antenna and converts them into a format that the computer can use.

Once the radio signal leaves the transmitter's antenna, it travels through the air and is picked up by receiving antennas. As the signal travels through the air, it loses its strength, eventually losing enough power that it cannot be accurately received.

Wireless networks take many forms. VHF radio, FM-AM radio, cellular phones, and CB radios are all forms of wireless technology but have very specific purposes (usually for the purpose of communicating verbal information). When we talk about wireless networking, it is about a breed of technology that is able to communicate *data*. Data can be voice, or Internet, or any other kind of computer information. This kind of wireless technology can be used to supplement or even replace existing wireless systems.

There are many wireless technologies suitable for data networking. When the concept of using radio signals to connect various computers in a building was introduced, the IEEE formed a committee to set the standards for the technology. That committee was called the 802.11 committee, and the various standards they developed are known as 802.11a, 802.11b, 802.11g, and so forth. This group of 802.11 standards became known as WiFi technology. Because WiFi technology quickly became popular, the cost of WiFi equipment has decreased rapidly. Many organizations and wireless Internet service providers (ISPs) have started with WiFi.

1.1.4 IEE 802.11 (WiFi)

WiFi is a common wireless technology used by home owners, small businesses, and starting ISPs. WiFi devices are available "off the shelf" from computer stores, and enhanced WiFi devices are designed for ISP use.

Advantages of WiFi are as follows:

- Ubiquitous and vendor neutral; any WiFi device will work with another regardless of the manufacturer

- Affordable cost
- Hackable; many "hacks" exist to extend the range and performance of a WiFi network

Disadvantages are as follows:

- It is designed for LANs, not wide area networking (WAN).
- It uses the CSMA mechanism. Only one wireless station can "talk" at a time, meaning one user can potentially hog all of the network's resources. Applications such as video conferencing, voice over Internet protocol (VOIP), and multimedia can take down a network.

1.1.5 IEE 802.16 (WiMAX)

WiMax is a superset of WiFi designed specifically for last-mile distribution and mobility. WiMax promises high speed (30+ Mbps). WiMax is a relatively new standard, so WiMax products are expensive.
    An advantage of WiMax is as follows:

- Specifically designed for wide area networking

Disadvantages of WiMax are the following:

- New technology that has not passed the test of time (yet)
- More expensive than WiFi

1.1.6 Hotspots

Hotspots are wireless networks often run by businesses and individuals. They are called "hotspots" because they provide a small coverage area for people to connect to community networks and the Internet; popular locations for hotspots include communal areas such as restaurants and cafes.
    Hotspots are also powerful tools for supporting tourism. Visitors to a hotspot can be presented with information about the local community, including upcoming events and even presentations of local artwork and artisan works. The BC Wireless Network Society of British Columbia, Canada, provides a service for a community wireless hotspot network.

1.1.6.1 Requirements to Use Wi-Fi Hotspots Computers (and other devices) connect to hotspots using a Wi-Fi network adapter. Newer laptops contain built-in adapters, but most other computers do not. Wi-Fi network adapters can be purchased and installed separately. Depending on the type of computer and personal preferences, USB, PC card, express card, or even PCI card adapters can be used.

Public Wi-Fi hotspots normally require a paid subscription. The sign-up process involves providing information regarding a credit card online or by phone and then choosing a service plan. Some service providers offer plans that are working at thousands of hotspots throughout the country.

Little technical information is also required to access Wi-Fi hotspots. The network name (SSID) distinguishes hotspot networks from each other. Encryption keys scramble the network traffic to and from a hotspot. Most businesses require these. Service providers supply this profile information for their hotspots.

1.1.6.2 Finding Wi-Fi Hotspots Computers can automatically scan for hotspots within range of their wireless signal. These scans identify the network name (SSID) of the hotspot, allowing the computer to initiate a connection.

Instead of using a computer to find hotspots, some people uses a separate gadget called a Wi-Fi finder. These small devices scan for hotspot signals similarly to computers, and many provide some indication of signal strength to help pinpoint their exact location. Before traveling to a faraway place, the location of Wi-Fi hotspots can be found using online wireless hotspot finder services.

1.1.6.3 Connection to Wi-Fi Hotspots The process for connecting to a Wi-Fi hotspot works similarly on home, business, and public wireless networks. With the profile (network name and encryption settings) applied on the wireless network adapter, initiation of connection is needed from the client's computer operating system (or software that was supplied with the network adapter). Paid or restricted hotspot services will require the user to log in with a user name and password at the first time of accessing the Internet.

1.1.6.4 Dangers of Wi-Fi Hotspots Although few incidents of hotspot security issues are reported, many people remain doubtful of their safety. Some caution is justified as a hacker with good technical skills can break into the computer through a hotspot and potentially access personal data. Taking a few basic precautions will ensure reasonable safety when using Wi-Fi hotspots:

1. Research the public hotspot service providers and choose only reputable ones who use strong security settings on their networks.
2. Ensure that you do not accidentally connect to nonpreferred hotspots by checking your computer's settings.
3. Finally, be aware of your surroundings and watch for suspicious individuals in the vicinity who may be reading your screen or even plotting to steal your computer.

Wi-Fi hotspots are now the common form of Internet access. Connecting to a hotspot requires a wireless network adapter, knowledge of the profile information of that hotspot, and sometimes a subscription to a paid service. Computers and Wi-Fi finder gadgets are capable of scanning the nearby area for Wi-Fi hotspots and several online services allow one to find faraway hotspot locations. Whether a home, business, or public hotspot is used, the connection process remains essentially the same. Likewise, as with any wireless network, security issues for Wi-Fi hotspots need to be managed.

A homeowner or renter likely has several options for how to connect to the Internet. The connection method affects how a home network must be set up to support Internet connection sharing.

1.1.7 Mesh Networking

Mesh networking is the holy grail of wireless networking. "Mesh" refers to many types of technology that enable wireless systems to find each other automatically and self-configure themselves to route information among themselves.

Mesh is as organic as networks can get, but it is very immature. Several implementations exist (but are not compatible with each other). Mesh networking should be treated as experimental, but community wireless networks make provisions for using mesh technology either during early deployment (where it may turn out to be stable for the needs of the community) or on an experimental basis. Most mesh products work under the Linux operating system and can use Prism 2.0 and 2.5 devices, or Atheros-based radios.

Some popular mesh protocols are as follows:

- AODV is an older protocol used by commercial and open source products such as LocustWorld. AODV appears to have many flaws and is not *necessarily* recommended.
- RoofNet is an experimental protocol from MIT that is being tested by community wireless networks throughout the world and appears to be very promising.

1.1.7.1 Limitation of Wireless Technology   The wireless radio spectrum is a finite resource. Many people can use the radio spectrum, but as more people use wireless networking, interference will increase. In some cases, you may even find your competitors actively working to interfere with you. It is important to adopt a policy early on in network deployment to work with the community to resolve interference issues. Network operators should inform each other when setting up a new wireless system. In fact, if licensed wireless devices are used, it is necessary to coordinate with other wireless users. Although coordination is not required when using license-exempt wireless devices, it is a best practice to follow.

1.2 Wireless Internet

Wireless Internet has become possible through the evolution of portable computers and wireless connections over a mobile telephone network. However, the realization of a mobile computing environment requires a communication architecture that not only is compatible with the current architectures but also takes into account the specific features of mobility and wirelessness.

In the last few years, we have seen an increase in the use of Internet systems as well as an increase in mobile communications. Now, many services of high utility to the end users are based on the Internet technology. If a convergence of the mobile and Internet technologies can be achieved, it would be powerful in realizing vast economies of scale as well as highly flexible service platforms. But, to manage a reliable wireless Internet, three kinds of constraints have to be studied:

- The wireless operating environment
- The existing Internet architecture
- The limitation of the end devices

Wireless networks are very interesting for

- Mobility
- Reduced installation time
- Increased reliability
- Long-term cost savings

The Internet is a cooperatively run collection of computer networks that span the globe. It is also a vast collection of resources: people, information, and multimedia. The word *Internet* describes a number of agreements, arrangements, and connections. In fact, it is a network of networks—more precisely, a network of LANs. Each individual network has its own domain and has specific resources and capabilities. Figure 1.1 shows a simple Internet connection.

The Internet offers a variety of services such as e-mail, keyboard-to-keyboard chatting, real-time voice and video communication, and transfer, storage, and retrieval of files. The Internet uses a system of packet

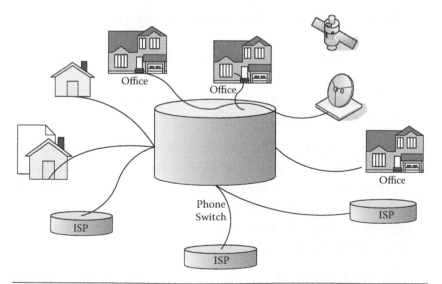

Figure 1.1   Internet connection.

switching for data transfer and was designed to be highly robust. In case one section of the network became inoperable, packets could simply be sent over another route and reach their destination. An important part of the IP protocol is the IP addressing standards, which define mechanisms to provide a unique address for each computer on the Internet. Users connect to an ISP via modems or integrated service digital networks (ISDNs) and the ISP routes the TCP/IP packets to and from the Internet.

The characteristics of wireless networks showed us that to manage reliable wireless Internet, we definitely have to consider the following subjects:

- Speed of wireless link
- Scalability
- Mobility
- Limited battery power
- Disconnection (voluntary or involuntary)
- Replication caching
- Handover

1.2.1 IP Limitations

The IP protocol has limitations due to the following characteristics:

- To send a packet on the Internet, a computer must have an IP address.
- This IP address is associated with the computer's physical location.
- TCP/IP protocol routes packets to their destination according to the IP address.

This leads to a big limitation. Indeed, within TCP/IP, if the mobile user moves without changing its IP address, the routing is lost; changing its IP address results in lost connections. In both cases, packets are lost. This leads to an unreliable network.

Regarding the specific features of mobility and wirelessness, wireless Internet must do the following:

- It should give mobile users the full Internet experience—not just a limited menu of specialized Web services or only e-mail.

- Voice telephony should migrate to the wireless Internet in due time.
- It should be reasonably fast—at least 100,000 bps throughput per user (about what has proved commercially successful over dial-up lines), with a growth path to millions of bits per second.
- It should work indoors and out, to both stationary and mobile users. (While drivers of vehicles should not be surfing the web, they may listen to Internet radio stations.)
- It should use power efficiently, since most devices will run on batteries or fuel cells for at least a few hours on a single charge.
- It should scale up to support millions of active devices, or more, within a single metropolitan region.

1.2.2 Mobile Internet Protocol (IP)

Mobile IP is an emerging set of protocols created by the Internet Engineering Task Force (IETF). Basically, it is a modification to IP that allows nodes to continue to receive packets independently of their connection point to the Internet. Figure 1.2 shows a mobile node communicating with other nodes after changing its link-layer point of attachment to the Internet; it does not change its IP address.

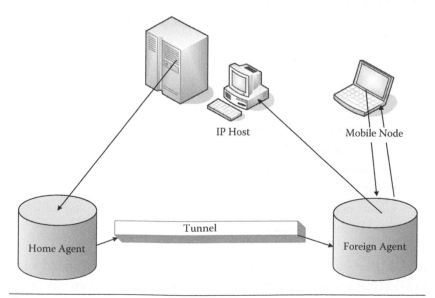

Figure 1.2   Datagram routing using mobile IP.

However, mobile IP is not suitable for fast mobility and smooth handover between cells, and a few requirements are to be considered for its design.

The messages used to transmit information about the location of a mobile node to another node must be authenticated to protect against remote redirection attacks.

For making the processes more secure and more efficient, enhancements to the mobile IP technique, such as mobile IPv6 and hierarchical mobile IPv6 (HMIPv6) defined in RFC 5380, are being developed to improve mobile communications. Researchers create support for mobile networking without requiring any predeployed infrastructure as is currently required by mobile IP. One such example is the interactive protocol for mobile networking (IPMN), which promises supporting mobility on a regular IP network just from the network edges by intelligent signaling between IP at endpoints and an application layer module with improved quality of service.

Researchers are also working to create support for mobile networking between entire subnets with support from mobile IPv6. One such example is the network mobility (NEMO) basic support protocol by the IETF Network Mobility Working Group, which supports mobility for entire mobile networks that move and attach to different points in the Internet. The protocol is an extension of mobile IPv6 and allows session continuity for every node in the mobile network as the network moves.

Changes in IPv6 for mobile IPv6 include the following:

- A set of mobility options to include in mobility messages
- A new home address option for the destination options header
- A new type 2 routing header
- New Internet control message protocol for IPv6 (ICMPv6) messages to discover the set of home agents and to obtain the prefix of the home link
- Changes to router discovery messages and options and additional neighbor discovery options

1.2.2.1 Working of Mobile IP   IP routes packets from a source endpoint to a destination by allowing routers to forward packets from incoming network interfaces to outbound interfaces according to information available in the routing tables. The routing tables typically maintain the next-hop information for each destination IP address according to the number of networks to which that IP address is connected. The network number is derived from the IP address by masking off some of the low-order bits. Thus, the IP address typically carries with it information that specifies the IP node's point of attachment.

To maintain existing transport-layer connections as the mobile node moves from place to place, it must keep its IP address the same. In TCP, connections are indexed by a quadruplet that contains the IP addresses and port numbers of both connection endpoints. Changing any of these four numbers will cause the connection to be disrupted and lost. On the other hand, correct delivery of packets to the mobile node's current point of attachment depends on the network number contained within the mobile node's IP address, which changes at new points of attachment. To change the routing requires a new IP address associated with the new point of attachment.

Mobile IP has been designed to solve this problem by allowing the mobile node to use two IP addresses. In mobile IP, the home address is static and is used, for instance, to identify TCP connections. The care-of address changes at each new point of attachment and can be thought of as the mobile node's topologically significant address; it indicates the network number and thus identifies the mobile node's point of attachment with respect to the network topology. The home address makes it appear that the mobile node is continually able to receive data on its home network; the mobile IP requires the existence of a network node known as the home agent. Whenever the mobile node is not attached to its home network (and is therefore attached to what is termed a foreign network), the home agent gets all the packets destined for the mobile node and arranges to deliver them to the mobile node's current point of attachment (see Figure 1.3).

Whenever the mobile node moves, it registers its new care-of address with its home agent. To get a packet to a mobile node from its home network, the home agent delivers the packet from the home network to the care-of address. The further delivery requires that the packet be modified so that the care-of address appears as the destination IP address. This modification can be understood as a packet transformation or, more specifically, a redirection. When the packet arrives at the care-of address, the reverse transformation is applied so

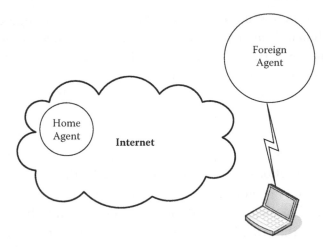

Figure 1.3   Mobile IP agents.

that the packet once again appears to have the mobile node's home address as the destination IP address. When the packet arrives at the mobile node, addressed to the home address, it will be processed properly by TCP or whatever higher level protocol logically receives it from the mobile node's IP (that is, layer 3) processing layer.

In mobile IP the home agent redirects packets from the home network to the care-of address by constructing a new IP header that contains the mobile node's care-of address as the destination IP address. This new header then shields or encapsulates the original packet, causing the mobile node's home address to have no effect on the encapsulated packet's routing until it arrives at the care-of address. Such encapsulation is also called tunneling, which suggests that the packet burrows through the Internet, bypassing the usual effects of IP routing.

A mobile node should minimize the number of administrative messages. Mobile IP must place no additional constraints on the assignment of IP addresses.

The mobile IP protocol can be described with the following steps:

Step 1: agent discovery
Step 2: registration home agent
Step 3: tunneling

A mobile node operating away from home registers its new care-of address with its home agent through the exchange of a registration request and registration reply messages. The home agent tunnels the information packets to the care-of address when the mobile node is away. Packets sent to the mobile node's home address are intercepted by its home agent, which tunnels them to the appropriate care-of address. There, the packets are delivered to the mobile node. In the reverse direction, packets sent by the mobile node may be delivered to their destination using a standard IP routing scheme, without necessarily passing through the home agent.

Mobile IP enables mobile computers to move about the Internet but remain addressable via their home network. Each mobile computer has an IP address (a home address) on its home network. Datagrams arriving for the mobile computer at its home network are subsequently repackaged for delivery to the mobile computer at its care-of address. The mobile computer informs its home agent about its current care-of address, using mobile IP registration protocols. When the home agent receives the mobile computer's care-of address, it binds that address with the mobile computer's home address, forming a binding that has an associated lifetime of validity.

This process is called registration, and it is transacted between the mobile computer and the home agent each time the mobile computer changes its point of attachment and receives a new care-of address. Often, the care-of address is advertised by an entity known as a foreign agent, which is located near the mobile computer and relays the registration messages back and forth between the mobile node and the home agent. Other times, the mobile computer itself acquires a care-of address by other means (notably, via the dynamic host configuration protocol [DHCP]) and assigns that care-of address to one of its own interfaces. This configuration is known as "colocated care-of address."

A mobile computer can easily switch between the two modes of operation depending upon the way in which care-of addresses are provided at its various points of attachment. Figure 1.4 shows a thumbnail sketch of a typical configuration, where the foreign agent has advertised the care-of address used by the mobile computer; the foreign agent and home agent are presumably and typically located on different subnets that have no a priori relationship to each other. If the mobile computer had attached via DHCP, there would be no foreign agent, but there would still be (typically) no relationship between the home network and the new point of attachment of the mobile computer.

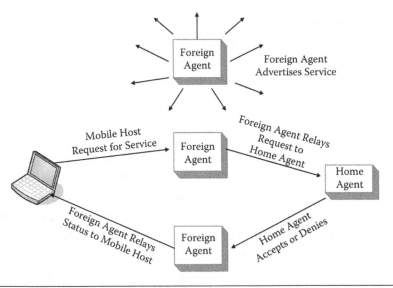

Figure 1.4 Registration operations in mobile IP.

When a home agent has a valid binding for the mobile node and a datagram for the mobile computer arrives at the home network, the home agent receives the datagram, acting as a proxy agent for the mobile computer on the home network. The home agent subsequently tunnels (by encapsulation) the datagram to the mobile computer's care-of address. The tunnel is the path between the home agent and the care-of address, and the care-of address is also known as the tunnel endpoint. After the datagram arrives at the tunnel endpoint, it is decapsulated and final delivery is made to the mobile computer. When the mobile node has a colocated care-of address, the final delivery is accomplished trivially.

Because traffic to the mobile node is controlled by correct operation of the mobile IP registration protocol, it is of essential importance that no corruption or intentional modifications of registration message data go undetected. If a malicious agent were able to register its own IP address as a false care-of address for the mobile node, the home agent would then route all the datagrams for the mobile node to the malicious agent instead. Clearly, the home agent must be able to ascertain that registration messages were issued authentically by the mobile node itself. This is accomplished by affixing a 128-bit digital signature, computed by using message-digest algorithm 5 (MD5) as a one-way hash function to the registration messages, and including protection against replay attacks in which a malicious node could record valid registrations for later replay, effectively disrupting the ability of the home agent to tunnel to the current care-of address of the mobile node at that later time.

1.2.3 Discovering the Care-of Address

The mobile IP discovery process has been built on top of an existing standard protocol: router advertisement. Mobile IP discovery does not modify the original fields of existing router advertisements but simply extends them to associate mobility functions. Thus, a router advertisement can carry information about default routers, just as before, and in addition carry further information about one or more care-of addresses. When the router advertisements are extended to contain the needed care-of address also, they are known as agent advertisements.

Home agents and foreign agents typically broadcast agent advertisements at regular intervals (for example, once a second or once every few seconds). If a mobile node needs to get a care-of address and does not wish to wait for the periodic advertisement, the mobile node can broadcast or multicast a solicitation that will be answered by any foreign agent or home agent that receives it. Home agents use agent advertisements to make themselves known, even if they do not offer any care-of addresses.

However, it is not possible to associate preferences to the various care-of addresses in the router advertisement, as is the case with default routers. The IETF working group was concerned that dynamic preference values might destabilize the operation of mobile IP. Because no one could defend static preference assignments except for backup mobility agents, which do not help distribute the routing load, the group eventually decided not to use the preference assignments with the care-of address list.

Thus, an agent advertisement performs the following functions:

- Allows for the detection of mobility agents
- Lists one or more available care-of addresses
- Informs the mobile node about special features provided by foreign agents—for example, alternative encapsulation techniques

- Lets mobile nodes determine the network number and status of their link to the Internet
- Lets the mobile node know whether the agent is a home agent, a foreign agent, or both and therefore whether it is on its home network or a foreign network

Mobile nodes use router solicitations to detect any change in the set of mobility agents available at the current point of attachment. (In mobile IP, this is then termed agent solicitation.) If advertisements are no longer detectable from a foreign agent that previously had offered a care-of address to the mobile node, the mobile node should presume that the foreign agent is no longer within range of the mobile node's network interface. In this situation, the mobile node should begin to hunt for a new care-of address, or possibly use a care-of address known from advertisements that it is still receiving. The mobile node may choose to wait for another advertisement if it has not received any recently advertised care-of addresses, or it may send an agent solicitation.

1.2.4 Registering the Care-of Address

Once a mobile node has a care-of address, its home agent must find out about it. Figure 1.4 shows the registration process defined by mobile IP for this purpose. The process begins when the mobile node, possibly with the assistance of a foreign agent, sends a registration request with the care-of address information. When the home agent receives this request, it (typically) adds the necessary information to its routing table, approves the request, and sends a registration reply back to the mobile node. Although the home agent is not required by the mobile IP protocol to handle registration requests by updating entries in its routing table, doing so offers a natural implementation strategy.

1.2.5 Authentication

Registration requests contain parameters and flags that characterize the tunnel through which the home agent will deliver packets to the care-of address. Tunnels can be constructed in various ways. When a home agent accepts the request, it begins to associate the home address of the mobile node with the care-of address, and maintains this association until the registration lifetime expires. The triplet that contains the home address, care-of address, and registration lifetime is called a "binding" for the mobile node. A registration request can be considered a "binding update" sent by the mobile node.

To secure the registration request, each request must contain unique data so that two different registrations will, in practical terms, never have the same MD5 hash. Otherwise, the protocol would be susceptible to replay attacks, in which a malicious node could record valid registrations for later replay, effectively disrupting the ability of the home agent to tunnel to the current care-of address of the mobile node at that later time. To ensure that this does not happen, mobile IP includes within the registration message a special identification field that changes with every new registration. The exact semantics of the identification field depend on several details, which are described at greater length in the protocol specification.

Briefly, there are two main ways to make the identification field unique. One is to use a time stamp; then, each new registration will have a later time stamp and thus differ from previous registrations. The other is to cause the identification to be a pseudorandom number; with enough bits of randomness, it is highly unlikely that two independently chosen values for the identification field will be the same. When randomness is used, mobile IP defines a method that protects both the registration request and reply from replay, and calls for 32 bits of randomness in the identification field. If the mobile node and the home agent get too far out of synchronization for the use of time stamps, or if they lose track of the expected random numbers, the home agent will reject the registration request and include information to allow resynchronization within the reply.

Using random numbers instead of time stamps avoids problems stemming from attacks on the network time protocol (NTP) that might cause the mobile node to lose time synchronization with the home agent or to issue authenticated registration requests for some future time that could be used by a malicious node to subvert a future registration. The identification field is also used by the foreign agent to match pending registration requests to registration replies when they arrive at the home agent and to be able subsequently to relay the reply to the mobile node.

The foreign agent also stores other information for pending registrations, including the mobile node's home address, the mobile node's media access control (MAC) layer address, the source port number for the registration request from the mobile node, the registration lifetime proposed by the mobile node, and the home agent's address. The foreign agent can limit registration lifetimes to a configurable value that it puts into its agent advertisements. The home agent can reduce the registration lifetime, which it includes as part of the registration reply, but it can never increase it.

1.2.6 Automatic Home Agent Discovery

When the mobile node cannot contact its home agent, Mobile IP has a mechanism that lets the mobile node try to register with another unknown home agent on its home network. This method of automatic home agent discovery works by using a broadcast IP address instead of the home agent's IP address as the target for the registration request. When the broadcast packet gets to the home network, other home agents on the network will send a rejection to the mobile node; however, their rejection notice will contain their address for the mobile node to use in a freshly attempted registration message. The broadcast is not an Internet-wide broadcast, but rather a *directed* broadcast that reaches only IP nodes on the home network.

1.2.7 Tunneling to the Care-of Address

Figure 1.5 shows the tunneling operations in mobile IP. The default encapsulation mechanism that must be supported by all mobility agents using mobile IP is IP-within-IP. Using IP-within-IP, the home agent, the tunnel source, inserts a new IP header, or tunnel header, in front of the IP header of any datagram addressed to the mobile node's home address. The new tunnel header uses the mobile node's care-of address as the destination IP address, or *tunnel destination*. The tunnel source IP address is the home agent, and the tunnel header uses 4 as the higher level protocol number, indicating that the next protocol header is again an IP header. In IP-within-IP, the entire original IP header is preserved as the first part of the payload of the tunnel header. Therefore, to recover the original packet, the foreign

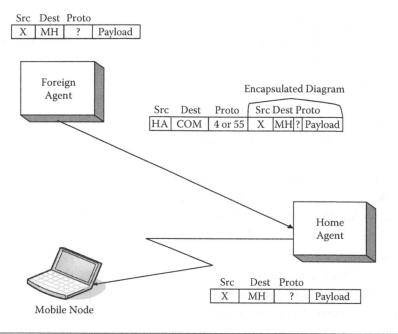

Figure 1.5 Tunneling operations in mobile IP.

agent merely has to eliminate the tunnel header and deliver the rest to the mobile node.

Figure 1.5 shows that sometimes the tunnel header uses protocol number 55 as the inner header. This happens when the home agent uses minimal encapsulation 11 instead of IP-within-IP. Processing for the minimal encapsulation header is slightly more complicated than that for IP-within-IP because some of the information from the tunnel header is combined with the information in the inner minimal encapsulation header to reconstitute the original IP header. On the other hand, header overhead is reduced.

1.2.8 Issues in Mobile IP

The most pressing outstanding problem facing mobile IP is that of security, but other technical as well as practical obstacles to deployment exist.

1.2.8.1 Routing Inefficiencies The base mobile IP specification has the effect of introducing a tunnel into the routing path followed by packets sent by the correspondent node to the mobile node. Packets from the mobile node, on the other hand, can go directly to the correspondent node with no tunneling required. This asymmetry is captured by the term *triangle routing*, where a single leg of the triangle goes from the mobile node to the correspondent node, and the home agent forms the third vertex controlling the path taken by data from the correspondent node to the mobile node.

1.2.8.2 Security Issues   A great deal of attention is being focused on making mobile IP coexist with the security features coming into use within the Internet. Firewalls, in particular, cause difficulty for mobile IP because they block all classes of incoming packets that do not meet specified criteria. Enterprise firewalls are typically configured to block packets from entering via the Internet that appear to emanate from internal computers. Although this permits management of internal Internet nodes without great attention to security, it presents difficulties for mobile nodes wishing to communicate with other nodes within their home enterprise networks. Such communications, originating from the mobile node, carry the mobile node's home address and would thus be blocked by the firewall. Mobile IP can be viewed as a protocol for establishing secure tunnels.

1.2.8.3 Ingress Filtering   Complications are also presented by ingress filtering operations. Many border routers discard packets coming from within the enterprise if the packets do not contain a source IP address configured for one of the enterprise's internal networks. Because mobile nodes would otherwise use their home address as the source IP address of the packets they transmit, this presents difficulty. Solutions to this problem in mobile IPv4 typically involve tunneling outgoing packets from the care-of address, but then the difficulty is how to find a suitable target for the tunneled packet from the mobile node. The only universally agreed on possibility is the home agent, but that target introduces yet another serious routing anomaly for communications between the mobile node and the rest of the Internet.

1.2.8.4 User Perceptions of Reliability   The design of mobile IP is founded on the premise that connections based on TCP should survive cell changes. However, opinion is not unanimous on the need for this feature. Many people believe that computer communications to laptop computers are sufficiently bursty that there is no need to increase the reliability of the connections supporting the communications. The analogy is made to fetching web pages by selecting the appropriate URLs (uniform resource locators). If a transfer fails, people are used to trying again. This is tantamount to making the user responsible for the retransmission protocol and depends for its acceptability on a widespread perception that computers and the Internet cannot be trusted to do things right the first time.

1.2.8.5 Issues in IP Addressing Mobile IP creates the perception that the mobile node is always attached to its home network. This forms the basis for the reachability of the mobile node at an IP address that can be conventionally associated with its fully qualified domain name (FQDN). If the FQDN is associated with one or more other IP addresses, perhaps dynamically, then those alternative IP addresses may deserve equal standing with the mobile node's home address. Moreover, it is possible that such an alternative IP address would offer a shorter routing path if, for instance, the address were apparently located on a physical link nearer to the mobile node's care-of address, or if the alternative address were the care-of address itself.

Finally, many communications are short-lived and depend on neither the actual identity of the mobile node nor its FQDN. Thus, they do not take advantage of the simplicity afforded by use of the mobile node's home address. These issues surrounding the mobile node's selection of an appropriate long-term (or not-so-long-term) address for use in establishing connections are complex and are far from being resolved.

1.2.8.6 Slow Growth in the Wireless Local Area Network (WLAN) Market Mobile IP has been engineered as a solution for WLAN location management and communications, but the WLAN market has been slow to develop. It is difficult to make general statements about the reasons for this slow development, but with the recent ratification of the IEEE 802.11 MAC protocol, WLANs may become more popular. Moreover, the bandwidth for wireless devices has been constantly improving, so radio and infrared devices on the market today offer multiple megabyte-per-second data rates. Faster wireless access over standardized MAC layers could be a major catalyst for growth of this market.

1.2.8.7 Competition from Other Protocols   Mobile IP may well face competition from alternative tunneling protocols such as PPTP and L2TP. These other protocols, based on point-to-point protocol (PPP), offer at least portability to mobile computers. If these alternative methods are made widely available, it is unclear if the use of mobile IP will be displaced or instead made more immediately desirable as people experience the convenience of mobile computing. In the future, it is also possible that mobile IP could specify use of such alternative tunneling protocols to capitalize on their deployment on platforms that do not support IP-within-IP encapsulation.

1.3 What Are Ad Hoc Networks?

An ad hoc network is a collection of wireless mobile nodes (or routers) dynamically forming a temporary network without the use of any existing network infrastructure or centralized administration. The routers are free to move randomly and organize themselves arbitrarily; thus, the network's wireless topology may change rapidly and unpredictably. Such a network may operate in a standalone fashion, or may be connected to the Internet. Multihop, mobility, and large network size combined with device heterogeneity, bandwidth, and battery power make the design of adequate routing protocols a major challenge. Some form of routing protocol is in general necessary in such an environment, since two hosts that may wish to exchange packets might not be able to communicate directly, as shown in Figure 1.6.

Mobile users will want to communicate in situations in which no fixed wired infrastructure is available. For example, a group of researcher's en route to a conference may meet at the airport and require connecting to the wide area network, students may need to interact during a lecture, or firemen need to connect to an ambulance en route to an emergency scene. In such situations, a collection of mobile hosts with wireless network interfaces may form a temporary network without the aid of any established infrastructure or centralized administration. Since nowadays many laptops are equipped with

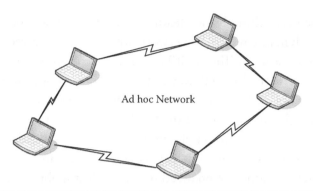

Figure 1.6 Mobile ad hoc networks.

powerful CPUs, large hard-disk drives, and good sound and image capabilities, the idea of forming a network among these researchers, students, or members of a rescue team, who can easily be equipped with devices mentioned before, seems possible. Such networks received considerable attention in recent years in both commercial and military applications, due to the attractive properties of building a network on the fly and not requiring any preplanned infrastructure such as base station or central controller.

A mobile ad hoc network (MANET) group has been formed within IETF. The primary focus of this working group is to develop and evolve MANET specifications and introduce them to the Internet standard track. The goal is to support mobile ad hoc networks with hundreds of routers and solve challenges in these kinds of networks. Some challenges that ad hoc networking is facing are limited wireless transmission range, hidden-terminal problems, packet losses due to transmission errors, mobility-induced route changes, and battery constraints.

Mobile ad hoc networks could enhance the service area of access networks and provide wireless connectivity into areas with previously poor or no coverage (e.g., cell edges). Connectivity to wired infrastructure will be provided through multiple gateways with possibly different capabilities and utilization. To improve performance, the mobile host should have the ability to adapt to variation in performance and coverage and to switch gateways when this would be beneficial. To enhance the prediction of the best overall performance, a network-layer metric has better overview of the network.

Ad hoc networking brings features like easy connection to access networks, dynamic multihop network structures, and direct peer-to-peer communication. The multihop property of an ad hoc network needs to be bridged by a gateway to the wired backbone. The gateway must have a network interface on both types of networks and be a part of both the global routing and the local ad hoc routing.

Users could benefit from ubiquitous networks in several ways. User mobility enables users to switch between devices, migrate sessions, and still get the same personalized services. Host mobility enables the users' devices to move around the networks and maintain connectivity and reachability.

1.3.1 Difference between Cellular and Ad Hoc Wireless Networks

Table 1.2 gives the major differences between cellular and ad hoc networks.

1.3.2 Applications of Ad Hoc Wireless Networks

The field of wireless networking emerges from the integration of personal computing, cellular technology, and the Internet. This is due to the increasing interactions between communication and computing, which is changing information access from "anytime, anywhere" to "all the time, everywhere." At present, a large variety of networks exists, ranging from the well known infrastructure of cellular networks to noninfrastructure wireless ad hoc networks.

Table 1.2   Differences between Cellular and Ad Hoc Wireless Networks

CELLULAR NETWORK	AD HOC WIRELESS NETWORK
Infrastructure network	Infrastructure-less network
Fixed, prelocated cell sites and base station	No base station and rapid deployment
Static backbone network topology	Highly dynamic network topologies with multihop
Relatively caring environment and stable connectivity	Hostile environment (noise, losses) and irregular connectivity
Detailed planning before base station can be installed	Ad hoc network automatically forms and adapts to changes
High setup costs	Cost effective
More setup time	Less setup time

The following are the applications of ad hoc wireless networks:

- Community network
- Enterprise network
- Home network
- Emergency response network
- Vehicle network
- Sensor network
- Education
- Entertainment
- Coverage extension
- Commercial and civilian environments

Unlike a fixed wireless network, wireless ad hoc or on-the-fly networks are characterized by the lack of infrastructure. Nodes in a mobile ad hoc network are free to move and organize themselves in an arbitrary fashion. Each user is free to roam about while communicating with others. The path between each pair of users may have multiple links, and the radio between them can be heterogeneous. This allows an association of various links to be a part of the same network. Mobile ad hoc networks can operate in a stand-alone fashion or could possibly be connected to a larger network such as the Internet.

Ad hoc networks are suited for use in situations where an infrastructure is unavailable or to deploy one is not cost effective. One of many possible uses of mobile ad hoc networks is in some business environments, where the need for collaborative computing might be more important outside the office environment than inside, such as in a business meeting outside the office to brief clients on a given assignment. Work has been going on to introduce the fundamental concepts of game theory and its applications in telecommunications. Game theory originates from economics and has been applied in various fields; it deals with multiperson decision making, in which each decision maker tries to maximize his utility. The cooperation of the users is necessary to the operation of ad hoc networks; therefore, game theory provides a good basis to analyze the networks.

A mobile ad hoc network can also be used to provide applications for crisis management services, such as in disaster recovery, where the entire communication infrastructure is destroyed and restoring communication quickly is crucial. By using a mobile ad hoc network, an infrastructure could be set up in hours instead of weeks, as is required in the case of wired-line communication. Another application example of a mobile ad hoc network is Bluetooth, which is designed to support a PAN by eliminating the need of wires between various devices, such as printers and personal digital assistants. The famous IEEE 802.11, or Wi-Fi protocol, also supports an ad hoc network system in the absence of a wireless access point.

The idea of ad hoc networking goes back to the US Defense Advanced Research Projects Agency (DARPA) packet radio network, which was in used in the 1970s. A mobile ad hoc network is a collection of mobile devices establishing a short-lived or temporary network in the absence of a supporting structure. Mobile ad hoc networks can be used in establishing efficient, dynamic communication for rescue, emergency, and military operations. A commercial application, such as Bluetooth, is one of the recent developments utilizing the concept of ad hoc networking.

Bluetooth is named after King Harald Blat (translated in English as King Harold Bluetooth), who ruled Denmark in the tenth century AD. Bluetooth was first introduced in 1998. It uses radio waves to transmit wireless data over short distances and can support many users in any environment. Eight devices can communicate with each other in a piconet. Ten of these piconets can coexist at one time in the same coverage range of the Bluetooth radio.

A Bluetooth device can act as both a client and a server. A connection must be established to exchange data between any two Bluetooth devices. To establish a connection, a device must request a connection with the other device. Bluetooth was based on the idea of advancing wireless interactions with various electronic devices. Devices like mobile phones, personal digital assistants, and laptops with the right chips could all communicate wirelessly with each other. However, it was later realized that a lot more is possible.

1.3.3 Technical and Research Challenges

Mobile ad hoc networks pose several technical and research challenges that need to be addressed. Ad hoc architecture has many benefits, such as self-reconfiguration and adaptability to high variable mobile characteristics such as power and transmission conditions, traffic distributions, and load balancing. These benefits pose new challenges. These mainly reside in the unpredictability to network topology due to mobility of nodes that, coupled with the local broadcast capability, causes a set of concerns in designing a communication system on top of ad hoc wireless networks. Many potential approaches have been proposed to deal with this issue: distributed MAC and dynamic routing, wireless service location protocol, wireless dynamic host configuration protocol, distributed admission call control, and quality of service (QoS)-based routing technique.

1.3.3.1 Security Issues and Challenges Security has become a primary concern in order to provide protected communication between mobile nodes in a hostile environment. Unlike the wire line networks, the unique characteristics of mobile ad hoc networks pose a number of nontrivial challenges to security design, such as open peer-to-peer network architecture, shared wireless medium, stringent resource constraints, and highly dynamic network topology. These challenges clearly make a case for building multifence security solutions that achieve both broad protection and desirable network performance.

LAYER	SECURITY ISSUES
Application layer	Detecting and preventing viruses, worms, malicious codes, and application abuses
Transport layer	Authenticating and securing end-to-end communications through data encryption
Network layer	Protecting the ad hoc routing and forwarding protocols
Link layer	Protecting the wireless MAC protocol and providing link-layer security support
Physical layer	Preventing signal jamming denial-of-service attacks

A fundamental vulnerability of MANETs comes from their open peer-to-peer architecture. Unlike wired networks that have dedicated routers, each mobile node in an ad hoc network may function as a router and forward packets for other nodes. The wireless channel is accessible to both legitimate network users and malicious attackers.

As a result, there is no clear line of defense in MANETs from the security design perspective. The boundary that separates the inside network from the outside world becomes blurred. There is no well defined place/infrastructure where we may deploy a single security solution. Moreover, portable devices, as well as the system security information they store, are vulnerable to compromises or physical capture, especially low-end devices with weak protection. Attackers may sneak into the network through these subverted nodes, which pose the weakest link and incur a domino effect of security breaches in the system.

The stringent resource constraints in MANETs constitute another nontrivial challenge to security design. The wireless channel is bandwidth constrained and shared among multiple networking entities. The computation capability of a mobile node is also constrained. For example, some low-end devices, such as PDAs, can hardly perform computation-intensive tasks like asymmetric cryptographic computation. Because mobile devices are typically powered by batteries, they may have very limited energy resources. The wireless medium and node mobility pose far more dynamics in MANETs compared to the wire-line networks. The network topology is highly dynamic as nodes frequently join or leave the network and roam in the network on their own will. The wireless channel is also subject to interferences and errors, exhibiting volatile characteristics in terms of bandwidth and delay. Despite such dynamics, mobile users may request for anytime, anywhere security services as they move from one place to another.

These characteristics of MANETs clearly make a case for building multifence security solutions that achieve both broad protection and desirable network performance. The security solution should

- Spread across many individual components and rely on their collective protection power to secure the entire network; the security scheme adopted by each device has to work within its own resource limitations in terms of computation capability, memory, communication capacity, and energy supply
- Span different layers of the protocol stack, with each layer contributing to a line of defense; no single-layer solution is possible to thwart all potential attacks

- Thwart threats from both outsiders, who launch attacks on the wireless channel and network topology, and insiders, who sneak into the system through compromised devices and gain access to certain system knowledge
- Encompass all three components of prevention, detection, and reaction that work in concert to guard the system from collapse
- Be practical and affordable in a highly dynamic and resource-constrained networking scenario

1.3.3.2 Different Types of Attacks on Multicast Routing Protocols

1.3.3.2.1 Rushing Attack Many demand-driven protocols such as on-demand multicast routing protocol (ODMRP), multicast ad hoc on-demand distance vector (MAODV), and adaptive demand-driven multicast routing protocol (ADMR), which use the duplicate suppression mechanism in their operations, are vulnerable to rushing attacks. When source nodes flood the network with route discovery packets to find routes to the destinations, each intermediate node processes only the first nonduplicate packet and discards any duplicate packets that arrive at a later time. Rushing attackers, by skipping some of the routing processes, can quickly forward these packets and be able to gain access to the forwarding group.

1.3.3.2.2 Black Hole Attack First, a black hole attacker needs to invade into the forwarding group—for example, by implementing rushing attack—to route data packets for some destination to itself. Then, instead of doing the forwarding task, the attacker simply drops all of the data packets that it receives. This type of attack often results in a very low packet delivery ratio.

1.3.3.2.3 Neighbor Attack Upon receiving a packet, an intermediate node records its ID in the packet before forwarding the packet to the next node. However, if an attacker simply forwards the packet without recording its ID in the packet, it makes two nodes that are not within the communication range of each other believe that they are neighbors (i.e., one hop away from each other), resulting in a disrupted route.

1.3.3.2.4 Jellyfish Attack   Similarly to the black hole attack, a jellyfish attacker first needs to intrude into the forwarding group and then it delays data packets unnecessarily for some amount of time before forwarding them. This results in significantly high end-to-end delay and delay jitter and thus degrades the performance of real-time applications.

1.3.3.3 Interconnection of Mobile Ad Hoc Networks and the Internet   The interconnection of mobile ad hoc networks to fixed IP networks is one of the topics receiving more attention within the MANET working group of the IETF as well as in many research projects funded by the European Union. Several solutions have recently been proposed, but at this time it is unclear which ones offer the best performance compared to the others. In addition to introducing the main challenges, design options that need to be considered are discussed in detail in this text.

1.3.4 Issues in Ad Hoc Wireless Networks

Different types of terminals form most of the ad hoc networks—for example, PDA-like devices, mobile phones, two-way pagers, sensors or desktop computers—with different capabilities in terms of maximum transmission power, energy availability, mobility patterns, and QoS requirements. Ad hoc networks are generally heterogeneous in terms of terminals and services offered. In terms of energy and power, one has to consider not only node heterogeneity in terms of transmission power and energy availability, but also varying communication ranges, such as sleeping or active modes and the existence of energy supplies. Ad hoc networks raise new issues concerning security and privacy.

Ad hoc networks inherit some of the traditional problems of wireless communication and wireless networking:

- The wireless medium does not have proper boundaries outside which nodes are known to be unable to receive network frames.
- The wireless channel is weak, unreliable, and unprotected from outside signals, which may cause lots of problems to the nodes in the network.
- The wireless channel has time-varying and asymmetric propagation properties.

• Hidden-nodes and exposed-nodes problems may occur.

1.3.4.1 Medium Access Control (MAC) Protocol Research Issues Wireless multiple access can be categorized into random access (e.g., CSMA, CSMA with collision detection [CSMA/CD]) and controlled access (e.g., TDMA and token-based schemes). Random access will be suitable for ad hoc networks because of lack of infrastructure support. In addition, as the basis for its standards, the IEEE 802.11 WLAN committee selected the CSMA/CA scheme. The Bluetooth technology that is designed to support beyond data traffic and delay sensitive applications (e.g., audio and video) adopted the TDMA scheme with an implicit token-passing scheme for the slots assignment. The use of Bluetooth and IEEE 802.11 is not optimized in multihop environments. These technologies are used for single hop wireless personal area networks (WPANs) and WLANs, respectively. The design of MAC protocols for a multihop ad hoc environment is a hot research issue.

1.3.4.2 Networking Issues Most of the main functionalities of the *networking protocols* need to be redesigned. Networking protocols uses one-hop transmission services provided by the enabling technologies to construct end-to-end delivery service from sender needs to locate the receiver inside the network. The purpose of the *location services* is to map to its current location in the network dynamically. Current solutions generally adopted to manage mobile terminals in infrastructure networks are inadequate and new approaches are to be found for mobile management.

A simple solution to node location is based on flooding the location query through the network. This approach is suitable only for limited size networks. Controlling the flooding area can help to refine the technique. This can be achieved by gradually increasing the number of hops involved in the flooding propagation until the node is located.

The flooding approach constitutes a reactive location service in which no location information is maintained inside the network. The location information service maintenance cost is negligible and all the complexity is associated with query operations. On the other hand, proactive location services subdivide the complexity in the two phases. Proactive services construct and maintain inside the network data structures that store the location information of each node. By exploiting the data structures, the query operations are highly simplified.

1.3.4.3 Ad Hoc Routing and Forwarding The highly dynamic nature of a mobile ad hoc network results in frequent and unpredictable changes of network topology, adding difficulty and complexity to routing among the mobile nodes. The challenges and complexities, coupled with the critical importance of routing protocol in establishing communications among mobile nodes, make routing area the most active research area with the MANET domain.

Numerous routing protocols and algorithms have been proposed. Their performance under various network environments and traffic conditions has been studied and compared. Several surveys and comparative analyses of MANET routing protocols have been published. The classification of the routing protocols can be done via the type of cast property—that is, whether they use a unicast, multicast, geocast, or broadcast routing protocol.

1.3.4.4 Unicast Routing A primary goal of unicast routing protocols is the correct and efficient route establishment and maintenance between a pair of nodes so that messages may be delivered reliably and in a timely manner. MANET characteristics make the direct use of these protocols infeasible. MANET routing protocols must operate in networks with highly dynamic topologies where routing algorithms run on resource-constrained devices.

MANET routing protocols are typically subdivided into two main categories: *proactive routing protocols* and *reactive protocols.* Proactive routing protocols are derived from distance-vector and link-state protocols. They maintain consistent and updated routing information for every pair of network nodes by proactively propagating route updates at fixed time intervals. As the routing information is usually maintained in tables, these protocols are also referred to as "table-driven protocols."

1.3.4.4.1 Proactive Routing Protocols "Proactive routing protocol" is the constant maintaining of a route by each node to all other network nodes. The route creation and maintenance are performed through both periodic and event-driven messages. The various proactive protocols are destination-sequenced distance-vector (DSDV), optimized link-state routing (OLSR), and topology dissemination based on reverse path forwarding (TBRPF).

The DSDV protocol is a distance-vector protocol with extensions to make it suitable to MANET. Every node maintains a routing table with one route recorded. To avoid routing loops, a destination sequence number is used. A node increments its sequence number whenever a change occurs in its neighborhood.

The OLSR protocol is an optimization for MANET of legacy link-state protocols. The key point of the optimization is the *multipoint relay* (MPR). By flooding a message to its MPRs, a node is guaranteed that the message, when retransmitted by the MPRs, will be received by all its two-hop neighbors. TBRPF is a link-state routing protocol that employs a different overhead reduction technique. Each node computes a shortest-path tree to all other nodes; however, to optimize bandwidth, only part of the tree is propagated to neighbors. The fisheye state routing (FSR) protocol is also an optimization over link-state algorithms using fisheye technique. FSR propagates link-state information to other nodes in the network based on how far away the nodes are.

1.3.4.4.2 Reactive Routing Protocols With these protocols, to reduce overhead, the route between two nodes is discovered only when it is needed. There are different types of reactive routing protocols such as dynamic source routing (DSR), ad hoc on-demand distance vector (AODV), temporally ordered routing algorithm (TORA), associatively based routing (ABR), and signal stability routing (SSR).

DSR is a loop-free, source-based, on-demand routing protocol where each node maintains a route cache that contains the source routes learned by the node. The route discovery process is initiated only when a source node does not already have a valid route to the destination in its route cache; entries in the route cache are continually updated as new routes are learned. Source routing is used for packet forwarding. AODV is another reactive improvement of the DSDV protocol. AODV minimizes the number of route broadcasts by creating routes on demand, as opposed to maintaining a complete list of routes as in the DSDV algorithm. Similarly to DSR, route discovery is initiated on demand, and the route request is then forwarded by the source to the destination.

TORA is another source-initiated, on-demand routing protocol built on the concept of link reversal of the directed acyclic graph (ACG). In addition to being loop free and bandwidth efficient, TORA has the property of being highly adaptive and quick in route repair during link adaptation and quick in route repair during link failure, while providing multiple routes for any desired source-destination pair.

The ABR protocol is also a loop-free protocol built using a new routing metric termed "degree of association stability" in selecting routes, so that the route discovered can be longer lived and thus more stable and requiring fewer subsequent updates. The limitation of ABR comes mainly from a periodic used to establish the association stability metrics, which may result in additional energy consumption. The signal stability algorithm (SSA) is basically an ABR protocol with the additional property of routes selection using the signal strength of the link.

1.3.4.4.3 Hybrid Protocols In addition to proactive and reactive protocols, another class of unicast routing protocols that can be identified is *hybrid protocols*. The zone-based hierarchical link-state routing protocol (ZRP) is an example of a hybrid protocol that combines both proactive and reactive approaches, thus trying to bring together the advantages of the two approaches. ZRP defines around each node a zone that contains the neighbors within a given number of hops from the node. Proactive and reactive algorithms are used by the node to route packets within and outside the zone, respectively.

1.3.4.4.4 Multicast Routing Multicasting is an efficient communication service for supporting multipoint applications. Two main approaches are used for multicast routing in fixed networks: group-shared tree and source-specific tree. In the group share, a single tree is constructed for the whole group. The source-specific approach maintains, for each source, a tree toward all its receivers. There are two types of multicast protocols: MAODV and ad hoc multicast routing protocol utilizing increasing ID numbers (AMRIS). Both are on-demand protocols and construct a shared delivery tree to support multiple senders and receivers within a multicast session.

The topology of a wireless mobile network can be very dynamic, and hence the maintenance of a connected multicast routing tree may cause large overheads. To avoid this, a different approach based on meshes has been proposed. Meshes are more suitable for dynamic environments because they support more connectivity than trees; thus, they support multicast trees. There are two types of mesh-based multicast routing protocols included: core-assisted mesh protocol (CAMP) and the on-demand multicast routing protocol (ODMRP). These protocols build routing meshes to disseminate multicast packets within groups. The difference is that ODMRP uses flooding to build the mesh, while CAMP uses one or more nodes to assist in building the mesh, instead of flooding.

1.3.4.5 Location-Aware Routing During forwarding operations, location-aware routing protocols use the nodes position provided by global positioning system (GPS) or other mechanisms. Specifically, a node selects the next hop for packets forwarding by using the physical position of its one-hop neighbors and the physical position of the destination node. Location-aware routing does not require router establishment and maintenance. No routing information is stored. The use of geo-location information avoids network-wide searches, as both control and data packets are sent toward the known geographical coordinates of the destination node.

Three main strategies can be identified in location-aware routing protocols: *greedy forwarding, directed flooding,* and *hierarchical routing:*

> *Greedy forwarding:* In this type of strategy, a node tries to forward the packet to one of its neighbors that is closer to the destination. If more than one node is closer, different choices are possible. If, on the other hand, no neighbor is closer, new rules are included in the greedy strategies to find an alternative route.
>
> *Direct flooding:* Directing flooding nodes forward the packets to all neighbors that are located in the direction of the destination. Distance routing effect algorithm for mobility

(DREAM) and location aid routing (LAR) are two routing algorithms that apply this principle.

Hierarchical routing: The location proxy routing protocol and the terminode routing protocol are hierarchical routing protocols in which routing is structured in two layers. Both protocols apply different rules to long- and short-distance routing, respectively. Location-aware routing is used for routing on long distances, while when a packet arrives close to the destination a proactive distance-vector scheme is adopted.

1.3.4.6 Transmission Control Protocol (TCP) Issues   TCP is an effective connection-oriented transport control protocol that provides the essential flow control and congestion control required to ensure reliable packet delivery. Numerous enhancements and optimizations have been proposed over the past few years to improve TCP performance for infrastructure-based WLANs and cellular networking environments. Infrastructure-based wireless networks are one-hop wireless networks where a mobile device uses the wireless medium to access the fixed infrastructure. The mobile multihop ad hoc environment brings fresh challenges to the TCP protocol.

The main research areas and open issues include the following:

- Impact of mobility
- Node interaction MAC layer
- Impact of TCP congestion window size
- Interaction between MAC protocols

1.3.4.7 Network Security   The wireless ad hoc nature of the MANET brings new security challenges to the network design. Wireless networks are generally more vulnerable to information and physical security threats than fixed wired networks. Vulnerability of channels and nodes, absence of infrastructure, and dynamically changing topology make ad hoc network security a difficult task. Broadcast wireless channels allow message eavesdropping and injection. The absence of infrastructure makes the classic security solutions based on certification authorities and online servers inapplicable. Routing the packets in a secured environment is another challenge.

1.3.4.8 Different Security Attacks Securing wireless ad hoc networks is a highly challenging issue. There are certain specific vulnerable attacks to the ad hoc context. Performing communication in free space exposes an ad hoc network and eavesdrop or inject messages. Ad hoc network attacks can be classified into active and passive attacks. A passive attack does not inject any message, but listens to the channel. A passive attack tries to discover valuable information and does not produce any new traffic in the network. In case of active attack, messages are inserted into the network; such attacks involve actions such as replication, modification, and deletion of exchanged data. In ad hoc networks, active attacks are impersonation, denial of service (DoS) and disclosure:

Impersonation: In this type, nodes may join the network unde-tectably or send false routing information, masquerading as some other trusted node. A black hole attack falls in this category: Here, a malicious node uses the routing protocol to advertise itself as having the shortest path to the node whose packets it wants to intercept.

Denial of service: Attacks like routing table overflow and sleep deprivation fall in this category.

Disclosure attack: A location disclosure attack can reveal something about the physical location of nodes or the structure of the network. Two types of security mechanism can be generally applied: preventive and detective. Preventive mechanisms are typically based on key-based cryptography. Key distribution is at the center of prevent mechanisms, since no central authority, no centralized trusted third party, and no central server are available ad hoc. Detective mechanisms have to monitor and rely on the audit trace that is limited to communication activities taking place within the radio range.

1.3.4.8.1 Attacks Using Fabrication In fabrication attacks, for disturbing the network operation or to consume other node resources, an intruder generates false routing messages, such as routing updates and route error messages. A number of fabrication-based attacks exist:

Resource consumption attack: In this attack, a malicious node deliberately tries to consume the resources (e.g., battery power, bandwidth, etc.) of other nodes in the network. The attacks could be in the form of unnecessary route request control messages, very frequent generation of beacon packets, or forwarding of stale information to nodes.

Rushing attack: On-demand routing protocols that use the route discovery process are vulnerable to this type of attack. An attacker node that receives a route request packet from the source node floods the packet quickly throughout the network before other nodes, which also receive the same route request packet, can react. Nodes that receive the legitimate route request packet assume those packets to be duplicates of the packet already received through the attacker node and hence discard those packets. Any route discovered by the source node would contain the attacker node as one of the intermediate nodes. Hence, the source node would not be able to find secure routes.

Black hole attack: Here, a malicious node falsely advertises a good path (e.g., shortest path or most stable path) to the destination node during the path-finding process. The intension of the malicious nodes could be to hamper the path-finding process or to interrupt all the data packets being sent to the concerned destination node.

Gray hole attack: The gray hole attack has two phases. In the first phase, a malicious node exploits the AODV protocol to advertise itself as having a valid route to a destination node with the intention of intercepting packets, even though the route is fake. In the second phase, the node drops the intercepted packets with a certain probability. This attack is more difficult to detect than the black hole attack where the malicious node drops the received data packets with certainty. A gray hole may exhibit its malicious behavior in different ways. It may drop packets coming from (or destined to) certain specific node(s) in the network while forwarding all the packets for other nodes. Another type of gray hole node may behave maliciously for some time duration by dropping packets but may switch to normal behavior later. A gray hole may also exhibit a behavior that is a combination of these two, thereby making its detection even more difficult.

Wormhole attack: In a wormhole attack, an attacker receives packets at one point in the network, forwards them to another point in the network, and then replays them into the network from that point. For tunneled distances longer than the normal wireless transmission range of a single hop, it is simple for the attacker to make the tunneled packet arrive with better metric than a normal multihop route. It is also possible for the attacker to forward each bit over the wormhole directly, without waiting for an entire packet to be received before beginning to tunnel the bits of the packet, in order to minimize delay introduced by the wormhole. Due to the nature of wireless transmission, the attacker can create a wormhole even for packets not addressed to itself, since it can overhear them in wireless transmission and tunnel them to the colluding attacker at the opposite end of the wormhole. If the attacker performs this tunneling honestly and reliably, no harm is done; the attacker actually provides a useful service in connecting the network more efficiently. The wormhole puts the attacker in a very powerful position relative to other nodes in the network, and the attacker could exploit this position in a variety of ways. The attack can also still be performed even if the network communication provides confidentiality and authenticity, and even if the attacker has no cryptographic keys. Furthermore, the attacker is invisible at higher layers; unlike a malicious node in a routing protocol, which can often easily be named, the presence of the wormhole and the two colluding attackers at either endpoint of the wormhole are not visible in the route.

1.3.4.9 Security at Data-Link Layer The wireless medium access protocol implements mechanisms based on cryptography to avoid unauthorized accesses and to enhance the privacy on radio links. The analysis on IEEE 802.11 and Bluetooth can be discussed in brief.

Security in the IEEE 802.11 standard is provided by the wired equivalent privacy (WEP) scheme, which supports both data encryption and integrity. Key is a 40-bit secret key that is shared by all the devices of a WLAN or is a pairwise secret key shared only by two communicating devices.

Bluetooth uses cryptographic security mechanisms implemented in the data-link layer. A key management service provides each device with a set of symmetric cryptographic keys required for the initialization of a secret channel with another device, the execution of an authentication protocol, and the exchange of encrypted data on the secret channel.

1.3.4.10 Secure Routing Malicious nodes can disrupt the correct functioning of a routing protocol by *modifying* routing information, *fabricating* false routing information, and *impersonating* other nodes. The secure routing protocol (SRP) is an extension that is applied to several existing reactive routing protocols. SRP is based on assumption of the existence of a security association between the sender and receiver based on a shared secret key negotiated at the connection setup. SRP fights against the attacks that disrupt the route discovery process. A node initiating a route discovery is able to identify and discard false routing information. Ariadne is a secure ad hoc routing protocol based on DSR and the timed efficient stream loss-tolerant authentication (TESLA) protocol.

The authenticated routing for ad hoc network (ARAN) protocol is an on-demand, secure, malicious action carried out by third parties in the ad hoc environment. ARAN is based on certificates received from a trusted certificate server before joining the ad hoc network.

Secure efficient ad hoc distance (SEAD) is a proactive secure routing protocol based on a routing table update message. The basic idea is to authenticate the sequence number and the metric field of a routing table update message using one-way hash functions. Hash chains and digital signature are used by the secure ad hoc on-demand distance vector (SAODV) mechanism.

1.3.4.11 Quality of Service (QoS) The ability of networks to provide QoS depends on the intrinsic characteristics of all the network components, from transmission links to MAC and network layers. Wireless links have a low and highly variable capacity, and high loss rates. Topologies are highly dynamic. Random access-based MAC protocols have no QoS support.

QoS on a MANET is not sufficient to provide a basic routing functionality. Other aspects that should also be taken into consideration are bandwidth constraints due to a shared media; dynamic topology, since nodes are mobile and the topology may change; and power consumption due to limited batteries.

For wired networks there are two approaches to obtain QoS: overprovisioning and network traffic engineering. Overprovisioning consists of the network operator offering a huge amount of resources such that the network can accommodate all the demanding applications. Network traffic engineering classifies ongoing connections and treats them according to a set of established rules. Two proposals belonging to this class have been done inside the IETF: (1) integrated services (IntServ) and (2) differentiated services (DiffServ).

IntServ is a reservation-oriented method where users request the QoS parameters they need. The resource reservation protocol (RSVP) has been proposed by IETF to set up resource reservations for IntServ. Opposite to IntServ, DiffServ is a reservation-less method. Using DiffServ, service providers offer a set of differentiated classes of QoS to their customers to support various types of applications. IPv4 TOS octet or the IPv6 traffic class octet is used to mark a packet to receive a particular QoS class.

In general, the wire-based QoS models are not appropriate for MANETs. Overprovisioning, for instance, may not be possible because resources are scarce. IntServ/RSVP may require unaffordable storage and processing for MNs (mobile nodes), and signaling overhead. Diffserv, on the other hand, is a lightweight overhead model that may be more suitable for MANETs. However, Diffserv organization in customers and service providers does not fit the distributed nature of MANETs. This has motivated numerous QoS proposals targeted to MANETs.

Quality of service for a network is measured in terms of guaranteed amount of data that a network transfers from one place to another in a given real-time slot, such as audio and video. This poses a number of different technical challenges.

The size of the ad hoc network is directly related to the quality of service of the network. If the size of the mobile ad hoc network is large, it might make the problem of network control extremely difficult. Communication between two participating nodes in mobile ad hoc networks can be seen as a complex end-to-end channel that changes routes with time.

In a mobile ad hoc network, a number of different routes with various levels of node capacity and power may be available for a source to transmit data to the destination. As a result, not all routes are capable of providing the same level of quality of service that can meet the requirements of mobile users. Moreover, even if the selected route between a source and the destination meets the user requirements, the network error characteristics are expected to vary with time due to the dynamic nature of mobile ad hoc networks.

Mobile ad hoc networks are expected to play an important role in the deployment of future wireless communication systems. Therefore, it is extremely important that these networks should be able to provide efficient quality of service that can meet the vendor requirements. To provide efficient quality of service in mobile ad hoc networks, there is a solid need to establish new architectures and services for routine network controls.

Variable link conditions are intrinsic characteristics in most mobile ad hoc networks. Rerouting among mobile nodes causes network topology and traffic load conditions to change dynamically. Given the nature of MANET, it is difficult to support real-time applications with appropriate QoS. In some cases it may be impossible to guarantee strict QoS requirements. But, at the same time, QoS is of great importance in MANETs since it can improve performance and allow critical information to flow even under difficult conditions.

Recent research activities have shown the importance of providing the QoS mechanism at multiple layers in the protocol stack. QoS-capable MACs and cross-layer design are emerging as potential solutions for QoS in MANET. QoS routing can be used by MANET routing protocols to select different paths to a destination depending on the packet characteristics.

QoS MAC protocols solve the problems of medium contention, support reliable unicast communications, and provide resource reservation for real-time traffic in a distributed wireless environment. Numerous MAC protocols and improvements have been proposed that can provide QoS guarantees to real-time traffic in a distributed wireless environment, including the GAMA/PR protocol and the black burst (BB) contention mechanism.

1.3.4.12 Simulation of Wireless Ad Hoc Networks Traditional modeling and simulation tools include NS2 (and recently NS3), OPNET modeler, and Jetsam. These tools focus primarily on the simulation of the entire protocol stack of the system. But the need for a more advanced simulation methodology is always there. Agent-based modeling and simulation offers this paradigm. Not to be confused with multiagent systems and intelligent agents, agent-based modeling originated from the social sciences, where the goal was to evaluate and view large-scale systems with numerous interacting "agents" or components in a wide variety of random situations to observe global phenomena. Unlike traditional AI systems with intelligent agents, agent-based modeling is similar to the real world. Agent-based models are thus effective in modeling biologically inspired and nature-inspired systems. In these systems, the basic interactions of the components in the system, also called a complex adaptive system, are simple but result in advanced global phenomena such as emergence.

Problems

1.1 Give the features of IrDA with suitable illustrations.

1.2 Compare and contrast Bluetooth with IrDA.

1.3 List the features of HomeRF.

1.4 Explain a typical wireless network with a suitable illustration.

1.5 Describe the advantages and disadvantages of WiFi and WiMax.

1.6 Discuss the technology deployed in wireless Internet.

1.7 Give the limitations of IP.

1.8 Discuss the working of datagram routing using mobile IP.

1.9 Explain the main issues involved in mobile IP.

1.10 What are ad hoc networks? Explain.

1.11 Differentiate between cellular and ad hoc wireless networks.

1.12 Give the applications of ad hoc networks.

1.13 Discuss in detail the technical and research challenges in ad hoc networks.

1.14 Describe the issues in ad hoc networks.

1.15 Explain the security problems in ad hoc networks.

1.16 Describe the features of Bluetooth.

1.17 What is Bluetooth?

1.18 Why is the technology called Bluetooth?

1.19 How is Bluetooth used?

1.20 What is a personal area network (PAN)?

1.21 Compare and contrast WiFi and WiMax.

1.22 Explain the functionalities of the WiFi hotspots.

1.23 How can WiFi hotspots be found?

1.24 What are the dangers of WiFi hotspots?

1.25 Explain the different types of attacks in ad hoc networks.

1.26 Explain the QoS issues in ad hoc networks.

1.27 Describe the different simulation tools used in ad hoc networks.

Bibliography

Barrett, C. et al. 2002. Characterizing the interaction between routing and MAC protocols in ad-hoc networks. *Proceedings of MobiHoc*, pp. 92–103.

Broch, J. et al. 1998. A performance comparison of multi-hop wireless ad hoc network routing protocols. *Proceedings of Mobicom*, pp. 85–97.

Frodigh, M. et al. 2000. Wireless ad hoc networking: The art of networking without a network. *Ericsson Review*, issue no. 4.

Haas, Z. J. et al., eds. 1999. Special issue on wireless ad hoc networks. *IEEE Journal on Selected Areas in Communications* 17 (8).

Larsson, T., and N. Hedman. 1998. Routing protocols in wireless ad-hoc networks—A simulation study. Master's thesis at Lulea University of Technology, Stockholm.

Ramanathan, R. 2001. Making ad hoc networks density adaptive. *Proceedings of MILCOM*, pp. 957–961.

Ramanathan, S., and M. Steenstrup. 1996. A survey of routing techniques for mobile communications networks. Baltzer/ACM *Mobile Networks and Applications* 1:89–104.

Ramasubramanian, V. et al. 2003. SHARP: A hybrid adaptive routing protocol for mobile ad hoc networks. *Proceedings of 4th International Symposium on Mobile Ad Hoc Networking and Computing (MobiHoc)*, Annapolis, MD, June 3–6.

Roux, N. et al. 2000. Cost adaptive mechanism to provide network diversity for MANET reactive routing protocols. *Proceedings of MILCOM*.

Royer, E. M., and C.-K. Toh. 1999. A review of current routing protocols for ad hoc mobile wireless networks. *IEEE Personal Communications* April: 46–55.

2
MAC LAYER PROTOCOLS

2.1 Introduction

The simplicity in deployment makes mobile ad hoc networks (which do not require any infrastructure) suitable for a variety of applications, such as collaborative computing, disaster recovery, and battle field communication. With the proliferation of communications and computing devices, such as mobile phones, laptops, or PDAs, personal area networking (PAN), which is an ad hoc networking-based technology, has recently gained much interest.

In ad hoc networks, transmitters use radio signals for communication. Generally, each node can only be a transmitter (TRX) or a receiver (RX), one at a time. Communication among mobile nodes is limited within a certain transmission range. And nodes share the same frequency domain to communicate. So, within such ranges, only one transmission channel is used, covering the entire bandwidth. Unlike wired networks, packet delay is caused not only by the traffic load at the node, but also by the traffic load at the neighboring nodes, which is called "traffic interference."

Medium access control (MAC) protocols play an important role in the performance of the mobile ad hoc networks (MANETs). A MAC protocol defines how each mobile unit can share the limited wireless bandwidth resource in an efficient manner. The source and destination could be far away and each time packets need to be relayed from one node to another in multihop fashion, a medium has to be accessed. Accessing media properly requires only informing the nodes within the vicinity of transmission. MAC protocols control access to the transmission medium. Their aim is to provide an orderly and efficient use of the common spectrum. These protocols are responsible for *per–link connection establishment* (i.e., acquiring the medium) and *per–link connection cancellation* (i.e., releasing the medium).

One of the fundamental challenges in MANET research is how to increase the overall network throughput while maintaining low energy consumption for packet processing and communications. The low throughput is attributed to the harsh characteristics of the radio channel combined with the contention-based nature of MAC protocols commonly used in MANETs.

Regarding the MAC protocol for a wireless mobile ad hoc network, the following performance measures should be considered:

* Throughput and delay: Throughput is generally measured as the percentage of successfully transmitted radio link level frames per unit time. Transmission delay is defined as the interval between the frame arrival time at the MAC layer of a transmitter and the time at which the transmitter realizes that the transmitted frame has been successfully received by the receiver.
* Fairness: Generally, fairness measures how fair the channel allocation is among the flows in the different mobile nodes. The node mobility and the unreliability of radio channels are the two main factors that impact fairness.
* Energy efficiency: Generally, energy efficiency is measured as the fraction of the useful energy consumption (for successful frame transmission) to the total energy spent.
* Multimedia support: This is the ability of a MAC protocol to accommodate traffic with different service requirements, such as throughput, delay, and frame loss rate.

2.2 Important Issues and Need for Medium Access Control (MAC) Protocols

There are several important issues in ad hoc wireless networks. Most ad hoc wireless network applications use the industrial, scientific, and medical (ISM) band, which is free of licensing formalities. Since wireless is a tightly controlled medium, it has limited channel bandwidth that is typically much less than that of wired networks. Also, the wireless medium is inherently error prone. Even though a radio may have sufficient channel bandwidth, factors such as multiple access, signal fading, and noise and interference can cause the effective throughput in wireless networks to be significantly lower. Since wireless nodes may be mobile, the network topology can change frequently without any predictable pattern. Usually the links between nodes would be bidirectional, but there may be cases when differences in transmission power give rise to unidirectional links, which necessitate special treatment by the MAC protocols.

Ad hoc network nodes must conserve energy as they mostly rely on batteries as their power source. The security issues should be considered in the overall network design, as it is relatively easy to eavesdrop on wireless transmission. Routing protocols require information about the current topology so that a route from a source to a destination may be found. However, the existing routing schemes, such as distance vector- and link state-based protocols, lead to poor route convergence and low throughput for dynamic topology. Therefore, a new set of routing schemes is needed in the ad hoc wireless context.

The MAC layer, sometimes also referred to as a sublayer of the data-link layer, involves the functions and procedures necessary to transfer data between two or more nodes of the network. It is the responsibility of the MAC layer to perform error correction for anomalies occurring in the physical layer. The layer performs specific activities for framing, physical addressing, and flow and error controls. It is responsible for resolving conflicts among different nodes for channel access. Since the MAC layer has a direct bearing on how reliably and efficiently data can be transmitted between two nodes along the routing path in the network, it affects the quality of service (QoS) of the network. The design of a MAC protocol should also address issues caused by mobility of nodes and an unreliable time-varying channel.

Design goals of the MAC protocol include the following:

- The operation of the protocol should be distributed.
- The protocol should provide QoS support for real-time traffic.
- The access delay, which refers to the average delay experienced by any packet to get transmitted, must be kept low.
- The available bandwidth must be utilized efficiently.
- The protocol should ensure fair allocation of bandwidth to nodes.
- Control overhead must be kept as low as possible.
- The protocol should minimize the effects of hidden and exposed terminal problems.

- The protocol must be scalable to large networks.
- The protocol should have power control mechanisms.
- The protocol should have mechanisms for adaptive data rate control.
- The protocol should try to use directional antennas.
- The protocol should provide synchronization among nodes.

2.2.1 Need for Special MAC Protocols

The popular carrier sense multiple access (CSMA) MAC scheme and its variations, such as CSMA with collision detection (CSMA/CD) developed for wired networks, cannot be used directly in the wireless networks, as explained later. In CSMA-based schemes, the transmitting node first senses the medium to check whether it is idle or busy. The node defers its own transmission to prevent a collision with the existing signal if the medium is busy. Otherwise, the node begins to transmit its data while continuing to sense the medium. However, collisions occur at receiving nodes. Since signal strength in the wireless medium fades in proportion to the square of distance from the transmitter, the presence of a signal at the receiver node may not be clearly detected at other sending terminals if they are out of range.

As illustrated in Figure 2.1, node B is within the range of nodes A and C, but A and C are not in each other's range. Let us consider the case where A is transmitting to B. Node C, being out of A's range, cannot detect a carrier and may therefore send data to B, thus causing a collision at B. This is referred to as the hidden-terminal problem, as nodes A and C are hidden from each other. Let us now consider another case where B is transmitting to A. Since C is within B's range, it senses a carrier and decides to defer its own transmission. However, this is unnecessary because there is no way that C's transmission can cause any collision at receiver A. This is referred to as the exposed-terminal problem, since B's being exposed to C caused the latter to defer its transmission needlessly. MAC schemes are designed to overcome these problems.

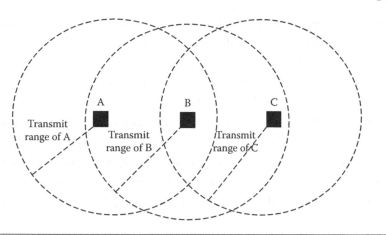

Figure 2.1 Illustration of hidden- and exposed-terminal problems.

2.3 Classification of MAC Protocols

This section describes the classification of MAC protocols and the various factors considered for classification. Various MAC schemes developed for wireless ad hoc networks can be classified as shown in Figure 2.2. In contention-free schemes (e.g., time division multiple access [TDMA], frequency division multiple access [FDMA], and code division multiple access [CDMA]), certain assignments are used to avoid contentions. Contention-based schemes, on the other hand, are aware of the risk of collisions of transmitted data. Since

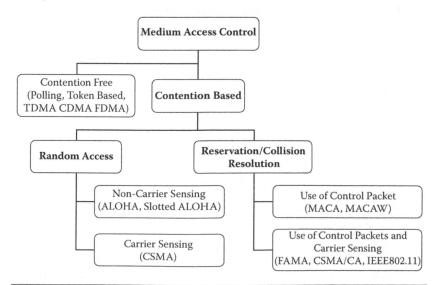

Figure 2.2 Classification of MAC protocols.

contention-free MAC schemes are more applicable to static networks and/or networks with centralized control, this chapter focuses on contention-based MAC schemes.

Another category depends on energy-efficient protocols at all layers of the network model. Hence, MAC protocols that are power aware are needed. Another class of MAC protocols uses directional antennas. The advantage of this method is that the signals are transmitted only in one direction. The nodes in other directions are therefore no longer prone to interference or collision effects, and spatial reuse is facilitated. Several MAC schemes have been proposed for unidirectional links.

Users will demand some level of QoS from MANET, such as end-to-end delay, available bandwidth, probability of packet loss, etc. However, the lack of centralized control, limited bandwidth channels, node mobility, power or computational constraints, and the error-prone nature of the wireless medium make it very difficult to provide effective QoS in ad hoc networks. Since the MAC layer has a direct bearing on how reliably and efficiently data can be transmitted from one node to the next along the routing path in the network, it affects the QoS of the network. Several QoS-aware MAC schemes are discussed in this chapter.

Another classification is based on the number of channels used for data transmission. Single-channel protocols set up reservations for transmissions and subsequently transmit their data using the same channel or frequency. Many MAC schemes use a single channel. Multiple-channel protocols use more than one channel in order to coordinate connection sessions among the transmitter and receiver nodes.

2.3.1 Contention-Based MAC Protocols

These protocols concentrate on the collisions of transmitted data. This includes two categories: random access and dynamic reservation/collision resolution protocols.

1. With random access-based schemes, such as ALOHA, a node may access the channel as soon as it is ready. Naturally, more than one node may transmit at the same time, causing

collisions. ALOHA is more suitable under low system loads with large numbers of potential senders and it offers relatively low throughput. A variation of ALOHA, termed slotted ALOHA, introduces synchronized transmission time slots similar to TDMA. In this case, nodes can transmit only at the beginning of a time slot. The introduction of time slots doubles the throughput as compared to the pure ALOHA scheme, with the cost of necessary time synchronization. The CSMA-based schemes further reduce the possibility of packet collisions and improve the throughput.

2. Dynamic reservation/collision resolution protocols: To solve the hidden- and exposed-terminal problems in CSMA, researchers have come up with many protocols that are contention based but involve some forms of dynamic reservation/collision resolution. Some schemes use the request-to-send/clear-to-send (RTS/CTS) control packets to prevent collisions (e.g., multiple access collision avoidance [MACA] and MACA for wireless LANs [MACAW]).

The contention-based MAC schemes can also be classified as sender-initiated versus receiver-initiated, single-channel versus multiple-channel, power-aware, directional antenna-based, unidirectional link-based, and QoS-aware schemes, as mentioned before. One distinguishing factor for MAC protocols is whether they rely on the sender initiating the data transfer or the receiver requesting the same. As mentioned previously, the dynamic reservation approach involves the setting up of some sort of a reservation prior to data transmission. If a node that wants to send data takes the initiative of setting up this reservation, the protocol is considered to be a sender-initiated protocol. Most schemes are sender initiated. In a receiver-initiated protocol, the receiving node polls a potential transmitting node for data. If the sending node indeed has some data for the receiver, it is allowed to transmit after being polled.

2.3.2 Contention-Based MAC Protocols with Reservation Mechanisms

The dynamic reservation approach involves the setting up of some sort of a reservation prior to data transmission. If a node that wants to send data takes the initiative of setting up this reservation, the protocol is considered to be a *sender-initiated protocol*. Most schemes are sender initiated. In a *receiver-initiated protocol*, the receiving node polls a potential transmitting node for data. If the sending node indeed has some data for the receiver, it is allowed to transmit after being polled. The MACA by invitation (MACA-BI) and receiver-initiated busy tone multiple access (RI-BTMA) are examples of such schemes.

2.3.2.1 Multiple Access Collision Avoidance (MACA)   The MACA protocol overcomes the hidden and exposed terminal problems. MACA uses two short signaling packets. The key idea of the MACA scheme is that any neighboring node that overhears an RTS packet has to defer its own transmissions until some time after the associated eight CTS packet would have finished, and that any node overhearing a CTS packet would defer for the length of the expected data transmission. In a hidden-terminal scenario, as explained in Section 2.2, *C* will not hear the RTS sent by *A*, but it would hear the CTS sent by *B*. Accordingly, *C* will defer its transmission during *A*'s data transmission. Similarly, in the exposed-terminal situation, *C* would hear the RTS sent by *B*, but not the CTS sent by *A*. Therefore, *C* will consider itself free to transmit during B's transmission.

It is apparent that this RTS-CTS exchange enables nearby nodes to reduce the collisions at the receiver, not the sender. Collisions can still occur between different RTS packets, though. If two RTS packets collide for any reason, each sending node waits for a randomly chosen interval before trying again. This process continues until one of the RTS transmissions elicits the desired CTS from the receiver. MACA is effective because RTS and CTS packets are significantly shorter than the actual data packets, and therefore collisions among them are less expensive compared to collisions among the longer data packets.

However, the RTS-CTS approach does not always solve the hidden terminal problem completely, and collisions can occur when different nodes send the RTS and the CTS packets. Let us consider an example with four nodes *A*, *B*, *C*, and *D* in Figure 2.3. Node *A* sends an RTS packet to *B*, and *B* sends a CTS packet back to *A*. At *C*, however, this CTS packet collides with an RTS packet sent by *D*. Therefore, *C* has no knowledge of the subsequent data transmission from *A* to *B*. While the data packet is being transmitted, *D* sends

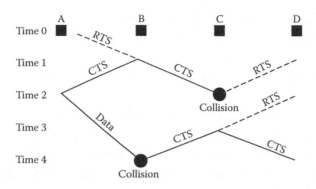

Figure 2.3 Illustration of failure of RTS-CTS mechanism in solving hidden- and exposed-terminal problems.

out another RTS because it did not receive a CTS packet in its first attempt. This time, *C* replies to *D* with a CTS packet that collides with the data packet at *B*. In fact, when hidden terminals are present and the network traffic is high, the performance of MACA degenerates to that of ALOHA.

Another *weakness* of MACA is that it does not provide any acknowledgment of data transmissions at the data-link layer. If a transmission fails for any reason, retransmission has to be initiated by the transport layer. This can cause significant delays in the transmission of data.

2.3.2.1.1 Applications for MACA If MACA proves effective, it may make single-frequency amateur packet radio networks practical. Although it would still be preferable for fixed backbones to use separate, dedicated channels or point-to-point links whenever possible, the ability to create usable, ad hoc, single-frequency networks could be very useful in certain situations. These include user access channels (such as 145.01 MHz in many areas) and in temporary portable and mobile operations where it is often infeasible to coordinate multifrequency networks in advance. This would be especially useful for emergency situations in remote areas without dedicated packet facilities.

2.3.2.1.2 Weaknesses of MACA
• When hidden terminals are present and the network traffic is high, the performance of MACA degenerates to that of ALOHA.

- MACA does not provide any acknowledgment of data transmissions at the data-link layer. If a transmission fails for any
  reason, retransmission has to be initiated by the transport layer.
  This can cause significant delays in the transmission of data.

2.3.2.1.3 Wireless (MACAW)   Multiple access with collision avoidance for wireless (MACAW) is a slotted MAC protocol widely used
in ad hoc networks. Furthermore, it is a foundation of many other
MAC protocols used in wireless sensor networks (WSNs). The IEEE
802.11 RTS/CTS mechanism is adopted from this protocol. It uses
the *RTS-CTS-DS-DATA-ACK* frame sequence for transferring data,
sometimes preceded by an *RTS-RRTS* frame sequence, in view to
provide a solution to the hidden-terminal problem. Although protocols based on MACAW, such as S-MAC, use carrier sense in addition to the RTS/CTS mechanism, MACAW does not make use of
carrier sense.

Principles of operation. Assume that node A has data to transfer
to node B. Node A initiates the process by sending a *request to send*
(RTS) frame to node B. The destination node (node B) replies with
a *clear to send* (CTS) frame. After receiving CTS, node A sends data.
After successful reception, node B replies with an acknowledgment
frame (ACK). If node A has to send more than one data fragment,
it has to wait a random time after each successful data transfer and
compete with adjacent nodes for the medium using the RTS/CTS
mechanism. Any node overhearing an RTS frame (for example, node
F or node E in the illustration) refrains from sending anything until
CTS is received or after waiting a certain time.

If the captured RTS is not followed by CTS, the maximum waiting
time is the RTS propagation time and the destination node turnaround
time. Any node overhearing a CTS frame refrains from sending anything for the time until the data frame and ACK should have been
received (solving the hidden-terminal problem), plus a random time.
Both the RTS and CTS frames contain information about the length
of the DATA frame. Hence, a node uses that information to estimate
the time for the data transmission completion. Before sending a long
DATA frame, node A sends a short data-sending (DS) frame, which
provides information about the length of the DATA frame. Every station that overhears this frame knows that the RTS/CTS exchange was

successful. An overhearing station, which might have received RTS and DS but not CTS, defers its transmissions until after the ACK frame should have been received plus a random time.

To sum up, a successful data transfer (A to B) consists of the following sequence of frames:

1. "Request to send" (RTS) frame from A to B
2. "Clear to send" (CTS) frame from B to A
3. "Data sending" (DS) frame from A to B
4. DATA fragment frame from A to B
5. Acknowledgment (ACK) frame from B to A

MACAW is a nonpersistent slotted protocol, meaning that after the medium has been busy—for example, after a CTS message—the station waits a random time after the start of a time slot before sending an RTS. This results in fair access to the medium. If, for example, nodes A, B, and C have data fragments to send after a busy period, they will have the same chance to access the medium since they are in transmission range of each other.

2.3.2.1.4 Floor Acquisition Multiple Access (FAMA) FAMA is another MACA-based scheme that requires every transmitting station to acquire control of the floor (i.e., the wireless channel) before it actually sends any data packet. Unlike MACA or MACAW, FAMA requires that collision avoidance be performed at both the sender and receiver nodes. To "acquire the floor," the sending node sends out an RTS using either nonpersistent packet sensing (NPS) or nonpersistent carrier sensing (NCS). The receiver responds with a CTS packet, which contains the address of the sending node. Any station overhearing this CTS packet knows about the station that has acquired the floor. The CTS packets are repeated long enough for the benefit of any hidden sender that did not register another sending node's RTS. The authors recommend the NCS variant for ad hoc networks since it addresses the hidden-terminal problem effectively.

2.3.2.2 IEEE 802.11 MAC Scheme The IEEE 802.11 specifies two modes of MAC protocol: distributed coordination function (DCF) mode (for ad hoc networks) and point coordination function (PCF) mode (for centrally coordinated infrastructure-based networks). The

DCF in IEEE 802.11 is based on CSMA with collision avoidance (CSMA/CA), which can be seen as a combination of the CSMA and MACA schemes. The protocol uses the RTS-CTS-DATA-ACK sequence for data transmission. The protocol not only uses physical carrier sensing, but also introduces the novel concept of virtual carrier sensing. This is implemented in the form of a network allocation vector (NAV), which is maintained by every node.

The NAV contains a time value that represents the duration up to which the wireless medium is expected to be busy because of transmissions by other nodes. Since every packet contains the duration information for the remainder of the message, every node overhearing a packet continuously updates its own NAV. Time slots are divided into multiple frames and there are several types of interframe spacing (IFS) slots. In increasing order of length, they are the short IFS (SIFS), point coordination function IFS (PIFS), DCF IFS (DIFS), and extended IFS (EIFS). The node waits for the medium to be free for a combination of these different times before it actually transmits. Different types of packets can require the medium to be free for a different number or type of IFS.

For instance, in ad hoc mode, if the medium is free after a node has waited for DIFS, it can transmit a queued packet. Otherwise, if the medium is still busy, a back-off timer is initiated. The initial back-off 10 value of the timer is chosen randomly from between 0 and CW-1, where contention window (CW) is the width of the contention window in terms of time-slots. After an unsuccessful transmission attempt, another back-off is performed with a doubled size of CW as decided by a binary exponential back-off (BEB) algorithm. Each time the medium is idle after DIFS, the timer is decremented. When the timer expires, the packet is transmitted. After each successful transmission, another random back-off (known as postback-off) is performed by the transmission-completing node. A control packet such as RTS, CTS, or ACK is transmitted after the medium has been free for SIFS.

2.3.2.3 Multiple Access Collision Avoidance by Invitation (MACA-BI)   In typical sender-initiated protocols, the sending node needs to switch to receive mode (to get CTS) immediately after transmitting the RTS. Each such exchange of control packets adds to turnaround time, reducing the overall throughput. MACA-BI [1] is a receiver-initiated protocol that reduces the number of such control packet exchanges. Instead of a sender waiting to gain access to the channel, MACA-BI requires a receiver to request the sender to send the data, by using a "ready-to-receive" (RTR) packet instead of the RTS and the CTS packets. Therefore, it is a two-way exchange (RTR-DATA) as against the three-way exchange (RTS-CTS-DATA) of MACA.

Because the transmitter cannot send any data before being asked by the receiver, there has to be a traffic prediction algorithm built into the receiver so that it can know when to request data from the sender. The efficiency of this algorithm determines the communication throughput of the system. The algorithm proposed by the authors piggybacks the information regarding packet queue length and data arrival rate at the sender in the data packet. When the receiver receives these data, it is able to predict the backlog in the transmitter and send further RTR packets accordingly. There is a provision for a transmitter to send an RTS packet if its input buffer overflows. In such a case, the system reverts to MACA. The MACA-BI scheme works efficiently in networks with predictable traffic patterns. However, if the traffic is bursty, the performance degrades to that of MACA.

2.3.2.4 Group Allocation Multiple Access with Packet Sensing (GAMA-PS) GAMA-PS incorporates features of contention-based as well as contention-free methods. It divides the wireless channel into a series of cycles. Every cycle is divided in two parts for contention and group transmission. Although the group transmission period is further divided into individual transmission periods, GAMA-PS does not require clock or time synchronization among different member nodes. Nodes wishing to make a reservation for access to the channel employ the RTS-CTS exchange. However, a node will back off only if it understands an entire packet. Carrier sensing alone is not sufficient reason for backing off.

GAMA-PS organizes nodes into transmission groups, which consist of nodes that have been allocated a transmission period. Every node in the group is expected to listen in on the channel. Therefore, there is no need for any centralized control. Every node in the group is aware of all the successful RTS-CTS exchanges and, by extension, of any idle transmission periods. Members of the transmission group take turns transmitting data, and every node is expected to send a begin transmission period (BTP) packet before actual data. The BTP contains the state of the transmission group, position of the node within that group, and the number of group members. A member station can transmit up to a fixed length of data, thereby increasing efficiency.

The last member of the transmission group broadcasts a transmit request (TR) packet after it sends its data. Use of the TR shortens the maximum length of the contention period by forcing any station that might contend for group membership to do so at the start of the contention period. GAMA-PS assumes that there are no hidden terminals. As a result, this scheme may not work well for mobile ad hoc networks. When there is not enough traffic in the network, GAMA-PS behaves almost like CSMA. However, as the load grows, it starts to mimic TDMA and allows every node to transmit once in every cycle.

2.3.3 MAC Protocols Using Directional Antennas

In the case of MAC protocols for ad hoc networks, omnidirectional antennas, which transmit and receive radio signals from all directions, are typically used. All other nodes in the vicinity remain silent. But the directional antennas attain higher gain and restrict the broadcast to a meticulous direction. Packet reception at a node with directional antennas is not exaggerated by intervention from other directions. Accordingly, depending on the direction of transmission, it is possible that two pairs of nodes located in each other's vicinity be in contact simultaneously. For the other untouched directions, this leads to a better spatial reuse. But these antennas' providing the correct direction and turning it into real time is not a trivial task.

Moreover, new protocols would need to be planned to take advantage of the new features provided by directional antennas because the current protocols (e.g., IEEE 802.11) cannot benefit from these features. Currently, directional antenna hardware is considerably bulkier and more expensive than omnidirectional antennas of comparable capabilities. Applications involving large military vehicles, however, are appropriate candidates for wireless devices using such antenna systems. The use of higher frequency bands (e.g., ultra-wideband transmission) will reduce the size of directional antennas. Many schemes have been proposed with this idea:

1. With packet radio networks and directional antennas, the slotted ALOHA scheme involves multiple directional antennas. In the context of beam-forming directional antennas, channel-access models, link power control, and directional neighbor discovery are required.

2. Through the use of special control packets, every node dynamically stores some information about its neighbors and their transmission schedules. This allows a node to guide its antenna appropriately based on the ongoing transmissions in the neighborhood. Using the directional antennas to apply a new form of link-state based routing is also proposed.

3. Using directional antennas, a directional MAC (D-MAC) scheme uses the familiar RTS-CTS-Data-ACK sequence where only the RTS packet is sent using a directional antenna. Though every node is assumed to be prepared with several directional antennas, only one of them is allowed to transmit at any given time, depending on the location of the intended receiver. Here, every node is aware of its own location as well as the locations of its direct neighbors. This scheme gives better throughput than IEEE 802.11 by allowing concurrent transmissions that are not possible in current MAC schemes.

4. In the IEEE 802.11 protocol, every node has multiple antennas. Any node that has data to send first sends out an RTS in all directions using every antenna. The intended receiver also sends out the CTS packet in all directions using all the antennas. The original sender is now able to discriminate which antenna picked up the strongest CTS signal and understands the relative direction of the receiver. The data packet is sent using the corresponding directional antenna in the direction of the intended receiver. Thus, the participating nodes need not know their location information in advance. Only one radio transceiver in a node can transmit and receive at any instant of time. This scheme can attain up to a two to three times better average throughput than CSMA/CA with the RTS/CTS scheme (using omnidirectional antennas).

5. Another scheme is multihop RTS MAC (M-MAC) for transmission on multihop paths. As directional antennas have a higher gain and transmission range than omnidirectional antennas, a node that is far away from another node communicates directly. For communicating with distant nodes, M-MAC uses multiple hops to send RTS packets to establish a link, but the subsequent CTS, data, and ACK packets are sent in a single hop. This protocol can achieve better throughput and end-to-end delay than the basic IEEE 802.11 and the D-MAC schemes. Depending on the topology configuration and flow patterns of the system, the performance also varies. Directional antennas bring in three new problems: (1) new kinds of hidden terminals, (2) higher directional interference (problems due to nodes that are in a straight line), and (3) deafness (where routes of two flows share a common link). These problems depend on the topology and flow patterns. With node mobility, the performance of these schemes will degrade. Some of the current protocols imprecisely assume that the gain of the directional antenna is the same as that of the omnidirectional antenna. But, none of them considers the effect of transmission power control, use of multiple channels, and support for real-time traffic.

2.3.4 Multiple-Channel MAC Protocols

The high probability of collision with the number of nodes is a major problem for the single shared channel scheme. This problem can be solved with multichannel approaches. As seen in the classification, some multichannel schemes use a dedicated channel for control packets (or signaling) and one separate channel for data transmissions. They set up busy tones on the control channel, one with small bandwidth consumption, so that nodes are conscious of the ongoing transmissions. Another approach is to use multiple channels for data packet transmission, which has the following advantages:

1. Use of a number of channels potentially increases the through-put, as the maximum throughput of a single channel scheme is limited by the bandwidth of that channel.
2. Transmissions of data on different channels do not get in the way of each other, and multiple transmissions can take place together in the same region. This leads to drastically fewer collisions.
3. It is easier to support QoS by using multiple channels.

In real time, a multiple data channel MAC protocol has to assign different channels to different nodes. The subject of medium access still needs to be resolved. This involves deciding, for instance, the time slots at which a node would get access to a particular chan-nel. In certain cases, it may be necessary for all the nodes to be harmonized with each other, whereas in other instances, it may be possible for the nodes to negotiate schedules among themselves. The details of some of the multiple channel MAC schemes are dis-cussed next.

2.3.4.1 Dual Busy Tone Multiple Access (DBTMA) In the schemes of exchange of RTS/CTS dialogue, these control packets themselves are prone to collisions. Thus, in the presence of hidden terminals, there remains a risk of data packet destruction due to collision. The DBTMA scheme effectively solves the hidden- and exposed-termi-nal problems by out-of-band signaling. Data transmission is happen-ing on the single shared wireless channel. It builds upon earlier work on the busy tone multiple access (BTMA) and the receiver-initiated busy tone multiple access (RI-BTMA) schemes. For managing access to the common medium, DBTMA decentralizes the responsibility and does not require time synchronization among the nodes.

Again, for setting up transmission requests, DBMTA sends RTS packets on data channels. Afterward, two different busy tones on a separate narrow channel are used to shield the transfer of the RTS and data packets. The sender of the RTS sets up a transmit-busy tone (BTt). Correspondingly, the receiver sets up a receive-busy tone (BTr) in order to acknowledge the RTS, without using any CTS packet. Any node that senses an existing BTr or BTt defers from sending its own RTS over the channel. Therefore, both of these busy tones together guarantee protection from collision from other nodes in the vicinity. Through the use of the BTt and BTr in combination, exposed terminals are able to initiate data packet transmissions. Simultaneously, hidden terminals can reply to RTS requests as data transmission occurs between the receiver and sender. However, the DBTMA scheme does not use ACK to acknowledge the received data packets. It requires supplementary hardware involvement.

2.3.4.2 Multichannel Carrier Sense Multiple Access (CSMA) MAC Protocol   The total available bandwidth (W) is divided into N distinct channels of W/N bandwidth each by multichannel CSMA protocol. Here, N is lower than the number of nodes in the network. Also, the channels are divided based on either an FDMA or CDMA scheme. A transmitter uses carrier sensing to see if the channel it last used is free or not. If found to be free, the channel used last is utilized. Otherwise, another free channel is chosen at random. If no free channel is found, the node should pull back and retry later.

Even when traffic load is very high and adequate channels are not accessible, chances of collisions are somewhat reduced since each node tends to prefer its last used channel instead of simply choosing a new channel arbitrarily. This protocol is more efficient than single-channel CSMA schemes. Interestingly, the performance of this scheme is lower than that of the single-channel CSMA scheme at lower traffic loads or when there are only a small number of active nodes for a long period of time. This happens due to the misuse of idling channels. The protocol is extended to select the best channel based on the signal power observed at the sender side.

2.3.4.3 Hop-Reservation Multiple Access (HRMA)   In ISM band, based on frequency hopping spread spectrum (FHSS) radios, HRMA is an efficient MAC protocol. For sending entire packets in the same hop, it uses time-slotting properties of very slow FHSS. For communication without intervention from other nodes, HRMA does not require carrier sensing; it employs a common frequency hopping sequence and also allows a pair of nodes to preserve a frequency hop (through the use of an RTS-CTS exchange). One of the N available frequencies in the network is reserved exclusively for synchronization. The remaining $N - 1$ frequencies are divided into

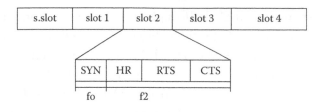

Figure 2.4   Structure of HRMA slot and frame.

$$M = \text{floor } ((N - 1)/2) \text{ pairs of frequencies}$$

For every pair, the first frequency is used for hop reservation (HR), RTS, CTS, and data packets, and the second frequency is used for ACK packets. HRMA is treated as a TDMA scheme, where time slots are assigned a specific frequency and subdivided into four parts: synchronizing, HR, RTS, and CTS periods. Figure 2.4 shows the HRMA frame. All through the synchronization phase of every time slot, all idle nodes synchronize to each other. On the other three periods, they jump together on the common frequency hops that have been assigned to the time slots.

First, the sender node sends an RTS packet to the receiver in the RTS period of the time slot. In that same time slot, the receiver sends a CTS packet to the sender in the CTS period. Now, the sender sends the data on the same frequency (at this time, the other idle nodes are synchronizing) and then hops to the acknowledgment frequency on which the receiver sends an ACK. If the data are large and require multiple time slots, the sender shows that in the header of the data packet. The receiver then sends an HR packet in the HR period of the next time slot, to lengthen the stipulation of the current frequency for the sender as well as the receiver. This indicates to the other nodes to omit this frequency in the hopping sequence. HRMA gets notably higher throughput than the slotted ALOHA in FHSS channels. It uses simple, half-duplex, slow-frequency hopping radios that are commercially available. On the other hand, it requires synchronization among the nodes, which is not proper for multihop networks.

2.3.4.4 Multichannel Medium Access Control (MMAC)   MMAC uses several channels by switching among them with dynamism. While the IEEE 802.11 protocol has intrinsic support for multiple channels in DCF mode, currently it utilizes only one channel. The prime cause is that hosts with a single half-duplex transceiver can only transmit or listen to one channel at a time. MMAC is a revision to the DCF in order to use multiple channels. Like the DPSM scheme, time is separated into multiple fixed-time beacon intervals. The commencement of every interval has a small ATIM window. In this window ATIM packets are exchanged between nodes for harmonization of the mission of appropriate channels to use in the subsequent time slots of that interval.

In contrast to other multichannel protocols, MMAC needs only one transceiver. At the commencement of every beacon interval, every node synchronizes itself to all other nodes by tuning into a common synchronization channel on which ATIM packets are exchanged. During this period of time no data packet diffusion is allowed. Moreover, every node keeps a preferred channel list (PCL) that stores the usage of channels within its transmission range and also allows for pointing priorities for those channels. When a node is sending a data packet, it sends out an ATIM packet to the receiver that includes the sender's PCL. The receiver then compares the sender's PCL with its own possession and selects a suitable channel to use. The response is given with an ATIM-ACK packet that includes the chosen channel in it.

If the chosen channel is up to standard for the sender, it responds with an ATIM-RES (reservation) packet. Any other node that overhears an ATIM-ACK or ATIM-RES packet updates its own PCL. Afterward, the sender and receiver swap RTS/CTS messages on the selected channel prior to data exchange. Otherwise, if the selected channel is not proper for the sender, it has to hang back till the next beacon interval tries another channel. In terms of throughput performance, MMAC is better than IEEE 802.11 and DCA. It can also be incorporated with IEEE 802.11 PSM mode using only simple hardware. Nonetheless, the packet delays are also longer than DCA. Furthermore, it is not appropriate for multihop ad hoc networks because the nodes are synchronized.

2.3.4.5 Dynamic   Channel   Assignment   with   Power   Control (DCA-PC) DCA-PC is an expansion of the DCA protocol that does not think about power control. It combines the concepts of power control and multiple-channel medium access in the framework of

MANETs. Dynamically, channels are assigned to the hosts, when they require them. The bandwidth is divided into a control channel and multiple data channels, and every node is prepared with two half-duplex transceivers. For exchanging control packets (using maximum power), one transceiver operates on the control channel for reserving the data channel; the other switches between the data channels for exchanging data and acknowledgments (with power control). When a host wants a channel to converse with another, it engages itself in an RTS/CTS/RES switchover, where RES is a special reservation packet signifying the appropriate data channel to be used.

Every node has a table of power levels for use while communication goes on with any other node. Based on the RTS/CTS exchanges on the control channel, these power levels are calculated. As entire node family always listen to the control channel, it can even update the power values with dynamism based on the other control exchanges going around it. Every node maintains a list of channel utilization information. This list informs the node which channel its neighbor is using and also the times of usage. DCA-PC has higher throughput than DCA. But, when the number of channels is increased beyond a point, the consequence of power control is considerably less for over-loading of the control channel. In summing up, DCA-PC is an original effort for solving dynamic channel assignment and power control issues in an integrated fashion.

2.3.5 Power-Aware or Energy-Efficient MAC Protocols

It is crucial to preserve energy and make use of power efficiently as mobile devices are battery power driven. Power preservation is considered across all the layers of the protocol stack. The following are the guiding principles for power conservation in MAC protocols:

1. Collisions should be avoided since they are the major basis of expensive retransmissions.
2. The transceivers consume most energy in active mode, so they should remain in standby mode (or switched off) when-ever possible.

3. The transmitter should be controlled to a lower power mode instead of maximum power because that is enough for the destination node to obtain the transmission.

The particulars of some chosen schemes are discussed next.

2.3.5.1 Power-Aware Medium Access Control with Signaling (PAMAS) The basic idea behind PAMAS is that all the RTS-CTS interactions are performed over the signaling channel and the data transmissions are kept apart over a data channel. The destination node starts sending out a busy tone over the signaling channel for receiving a data packet. When the power is down, nodes start listening to the signaling channel by assuming that it is most advantageous for them to power down their transceivers. For assuring that there are no power drops, every node makes its own decision whether or not to power off so that there is no drop in the throughput. When a node has nothing to transmit, it powers itself off and realizes that its neighbor is transmitting. A node also powers itself off if at least one neighbor is transmitting and another is receiving at the same time.

There are quite a few rules to decide the span of a power-down state. This scheme is used with other protocols like FAMA. The use of ACK and transmission of multiple packets improves the performance of PAMAS. The radio transceiver turnaround time, which is not negligible, is measured in the PAMAS scheme.

2.3.5.2 Dynamic Power-Saving Mechanism (DPSM) The concept of sleep and wake states of nodes is used in DPSM for preserving power. It is a type of the IEEE 802.11 scheme where it uses dynamically sized ad hoc traffic indication message (ATIM) windows for achieving longer dozing times of nodes. The IEEE 802.11 DCF mode is a power-saving mechanism where time is divided into beacon intervals used to synchronize nodes. At the commencement of each beacon interval, all nodes must stay wakeful for a fixed period of time known as the ATIM window, which announces the status of packets ready for transmission to the receiver nodes. These announcements are made through ATIM frames and are acknowledged via ATIM-ACK packets during the same beacon interval. Figure 2.5 shows the method. Performance suffers in

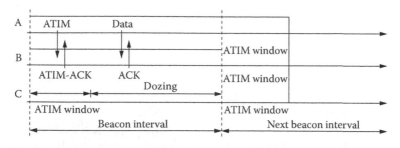

Figure 2.5 Power-saving mechanism for DCF.

terms of throughput and energy consumption if the size of the ATIM window is kept fixed.

Each node dynamically and independently chooses the length of the ATIM window in DPSM. As a consequence, every node ends up with a differently sized window. After participating in the transmission of packets announced in the prior ATIM frame, it allows the sender and receiver nodes to go into sleep mode instantly. Contrasting to the DCF method, they do not stay awake for the whole beacon interval.

After the current window expires, the length of the ATIM window is enlarged if some packets queued in the outgoing buffer have still not been sent. Again, each data packet carries the current length of the ATIM window and any nodes that eavesdrop on this information may decide to alter their own window length based on the received information. In terms of power saving and throughput, DPSM is more efficient than IEEE 802.11 DCF. Nonetheless, IEEE 802.11 and DPSM are not appropriate for multihop ad hoc networks as they take for granted that the clocks of the nodes are synchronized and the network is connected. Three variations of DPSM for multihop MANETs are available that use asynchronous clocks.

Node A announces a buffered packet for B using an ATIM frame. Node B replies by sending an ATIM-ACK and both A and B stay awake during the entire beacon interval. The actual data transmission from A to B is completed during the beacon interval; since C does not have any packet to send or receive, it dozes after the ATIM window.

2.3.5.3 Power-Control Medium Access Control (PCM) Alternating sleep and wake states for nodes were used in the previous concepts for power control. In PCM, the RTS and CTS packets are sent by means of the maximum power on hand, while the data and ACK

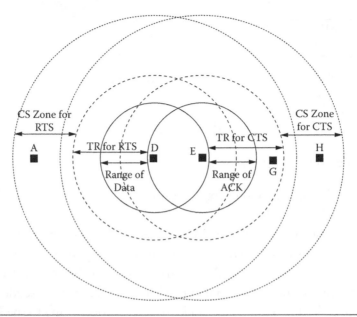

Figure 2.6   Illustration of power control scheme: CS = carrier sense and TR = transmission range.

packets are sent with the minimum power needed for communication. An example is shown in Figure 2.6. Node D sends the RTS to node E at transmit power level P_{max} and includes this value in the packet. E measures the genuine signal strength P_r of the received RTS packet. Founded on P_{max}, P_r and the noise intensity at its location, E then calculates the minimum power level, P_{suff}, that is truly enough for use by D. Now, when E responds with the CTS packet using the maximum power it has, it includes P_{suff} that D consequently uses for data transmission. G is capable of hearing this CTS packet and defers its own transmissions.

E also includes the power level, which it uses for the transmission in the CTS packet. D then follows a similar process and computes the least required power level that would get a packet from E to itself. It includes this value in the data packet so that E can use it to send the ACK. PCM also stipulates that the source node transmit the DATA packet from time to time at the maximum power level so that nodes in the carrier sensing range, such as A, may sense it. PCM thus gets energy savings without any throughput deprivation. The operation of the PCM scheme needs an accurate assessment of the signal strength of the received packet. Thus, the dynamics of wireless signal spread due to fading and shadowing effects degrade its performance. Another downside of this scheme is the complications in implementing frequent changes in the transmit power levels.

2.3.5.4 Power-Controlled Multiple Access (PCMA) PCMA relies on controlling transmission power of the sender. Then the intended receiver is able to decipher the packet. Interference with other neighboring nodes that are not involved in the packet exchange is avoided. PCMA uses two channels, one for sending busy tones and the other for data and control packets. Power control method in PCMA is used to increase channel efficiency all the way through spatial frequency reuse rather than only rising battery life. As a result, an important concern for the transmitter and receiver pair is to settle on the minimum power level required for the receiver to decode the packet, while distinguishing it from noise/interference. So as not to trouble the ongoing reception of the receiver by any potential transmitter, the receiver advertises its noise tolerances.

In the conventional methods for collision avoidance, a node is either permitted for transmission or not, depending upon the outcome of carrier sensing. In PCMA, this method is generalized to a surrounded power model. The sender sends a Request_Power_To_ Send (RPTS) packet in the data channel to the receiver before the transmission of data. The receiver responds with an Accept_Power_To_Send (APTS) packet in the data channel. This RPTS-APTS switching is used for the determination of minimum transmission power level, which causes a successful packet reception at the receiver. After this exchange, the genuine data are transmitted and acknowledged with an ACK packet.

Every receiver sets up a special busy tone as a periodic pulse in separate channels. The signal power of this busy tone advertises to the other nodes the added noise power the receiver node can bear. When a sender monitors the busy-tone channel, it is doing something similar to carrier sensing, as in CSMA/CA model. When a receiver sends out a busy-tone pulse, it is doing something similar to sending out a CTS packet. The RPTS-APTS exchange is equivalent to the RTS-CTS exchange. But the major difference is that the RPTS-APTS exchange does not force other hidden transmitters to back off. Collisions are resolved by the use of an appropriate back-off policy.

2.4 Summary

This chapter has focused on ad hoc wireless networks with respect to MAC protocols. Many schemes and their prominent features were discussed. Particularly, it has concentrated on the issues of collision resolution, power conservation, multiple channels, and advantages of using directional antennas. The individuality and operating principles of several MAC schemes were discussed. While some of them are general-purpose protocols (such as MACA, MACAW, etc.), others focus on specific features such as power control (PAMAS, PCM, etc.) or the use of dedicated technology like directional antennas (D-MAC, multihop RTS MAC, etc.). Most of these schemes, however, are not designed especially for networks with mobile nodes. On the contrary, the transaction time at the MAC layer is moderately short. The consequence of mobility will become less significant as the available channel bandwidth continues to grow.

Problems

2.1 Explain the need of MAC protocols for ad hoc networks.

2.2 Discuss the various factors that need to be considered while measuring the performance of the MAC protocol for ad hoc networks.

2.3 With a neat diagram, explain hidden- and exposed-node problems.

2.4 Give the classification of MAC protocols with a suitable example.

2.5 Describe contention-based MAC protocols with an example.

2.6 Explain contention-based MAC protocols with reservations.

2.7 Describe the MACA protocol with a suitable example.

2.8 Explain briefly the two modes of operation in the IEEE 802.11 MAC scheme.

2.9 Discuss the operation of multiple access collision avoidance by invitation (MACA-BI) with an illustration.

2.10 How would group allocation multiple access with packet sensing (GAMA-PS) work well with ad hoc networks? Explain.

2.11 Describe MAC protocols using directional antennas in detail.

2.12 Explain the advantage of multiple-channel MAC protocols.

2.13 Discuss the advantage of power-aware or energy-efficient MAC protocols.

2.14 Explain power-aware medium access control with signaling (PAMAS) with an example.

2.15 Explain how the dynamic power-saving mechanism (DPSM) is an efficient MAC protocol in ad hoc networks.

2.16 Discuss power-control medium access control (PCM) and power-controlled multiple access (PCMA).

2.17 What are the design goals of the MAC protocol?

2.18 Describe the applications of the MACA protocol.

2.19 Discuss the weaknesses of the MACA protocol.

2.20 Explain the principle of operation of the MACAW protocol.

Reference

1. Talucci, F., and M. Gerla. 1997. MACA-BI (MACA by invitation). A wireless MAC protocol for high speed ad hoc networking. *Proceedings of IEEE ICUPC.*

Bibliography

Abramson, N. 1970. The ALOHA system—Another alternative for computer communications. *Proceedings of Fall Joint Computer Conference* 37:281–285.

Akyildiz, I. F., J. McNair, L.C. Martorell, R. Puigjaner, and Y. Yesha. 1999. Medium access control protocols for multimedia traffic in wireless networks. *IEEE Network* 13:39–47.

Bhargavan, V., A. Demers, S. Shenker, and L. Zhang. 1994. MACAW: A media access protocol for wireless LANs. *Proceedings of ACM SIGCOMM,* 212–225.

Chakrabarti, S., and A. Mishra. 2001. QoS issues in ad hoc wireless networks. *IEEE Communications Magazine* 39 (2): 142–148.

Chen, B., K. Jamieson, H. Balakrishnan, and R. Morris. 2001. Span: An energy-efficient coordination algorithm for topology maintenance in ad hoc wireless networks. *ACM MOBICOM,* July.

Chen, K. C. 1994. Medium access protocols of wireless LANs for mobile computing. *IEEE Network* 8 (5): 50–63.

Chockalingam, A., and M. Zorzi. 1998. Energy efficiency of media access protocols for mobile data networks. *IEEE Transactions on Communications* 46 (11): 1418–1421.

Crow, B. P., I. Widjaja, J. G. Kim, and P. T. Sakai. 1997. IEEE 802.11 wireless local area networks. *IEEE Communications Magazine*: 116–126.

Fullmer, C. L., and J. J. Garcia-Luna-Aceves. 1995. Floor acquisition multiple access (FAMA) for packet-radio networks. *Proceedings of ACM SIGCOMM,* Cambridge, MA, Aug. 28–Sep. 1.

Gallager, R. G. 1985. A perspective on multi access channels. *IEEE Transactions on Information Theory* 31 (2): 124–142.

Goldsmith, A. J., and S. B. Wicker. 2002. Design challenges for energy-constrained ad hoc wireless networks. *IEEE Wireless Communications* 9 (4): 8–27.

Haas, Z. J., and J. Deng. 2002. Dual busy tone multiple access (DBTMA)—A multiple access control scheme for ad hoc networks. *IEEE Transactions on Communications* 50 (6): 975–984.

Haas, Z. J., and S. Tabrizi. 1998. On some challenges and design choices in ad hoc communications. *Proceedings of IEEE MILCOM98.* Vol. 1.

IEEE 802.11 Working Group. 1997. Wireless LAN medium access control (MAC) and physical layer (PHY) specification.

IETF MANET Working Group. http://www.ietf.org/html.charters/manet-charter.html

Jung, E.-S., and N. H. Vaidya. 2002. An energy-efficient MAC protocol for wireless LANs. *IEEE INFOCOM.*

———. 2002. A power control MAC protocol for ad hoc networks. *ACM International Conference on Mobile Computing and Networking (MOBICOM),* Sept.

Karn, P. 1990. MACA—A new channel access method for packet radio. *ARRL/CRRL Amateur Radio 9th Computer Networking Conference,* Sept. 22.

Kleinrock, L., and F. A. Tobagi. 1975. Packet switching in radio channels: Part I—Carrier sense multiple access modes and their throughput-delay characteristics. *IEEE Transactions on Communications* 23:1400–1416.

———. 1975. Packet switching in radio channels: Part II—The hidden terminal problem in carrier sense multiple access and busy tone solution. *IEEE Transactions on Communications* 23:1417–1433.

Monks, J., V. Bharghavan, and W. Hwu. 2001. A power controlled multiple access protocol for wireless packet networks. *IEEE INFOCOM,* April.

Muir, A., and J. J. Garcia-Luna-Aceves. 1998. An efficient packet sensing MAC protocol for wireless networks. *Mobile Networks and Applications* 3 (3): 221–234.

O'Hara, B., and Petrick, A. 1999. *IEEE 802.11 handbook: A designer's companion.* New York: IEEE Press.

Perkins, C. E. 2001. *Ad hoc networking.* Boston, MA: Addison-Wesley.

Poojary, N., S. V. Krishnamurthy, and S. Dao. 2001. Medium access control in a network of ad hoc mobile nodes with heterogeneous power capabilities. *Proceedings of IEEE ICC* 3:872–877.

Ramanathan, R., and R. R. Hain. 2000. Topology control of multihop wireless networks using transmit power adjustment. *IEEE INFOCOM* 2:404–413.

Rodoplu, V., and T. H. Meng. 1999. Minimum energy mobile wireless networks. *IEEE Journal on Selected Areas in Communication* 17 (8): 1338–1344.

Royer, E. M., and C. K. Toh. 1999. A review of current routing protocols for ad hoc mobile wireless networks. *IEEE Personal Communications* 6 (2): 46–55.

Sidhu, G., R. Andrews, and A. Oppenheimer. 1989. *Inside AppleTalk*. Boston, MA: Addison-Wesley.

Singh, S., and C. S. Raghavendra. 1998. PAMAS—Power aware multi-access protocol with signaling for ad hoc networks. *ACM Computer Communications Review* 28 (3): 5–26.

Sivalingam, K. M., M. B. Srivastava, and P. Agrawal. 1997. Low-power link and access protocols for wireless multimedia networks. *IEEE VTC*, May.

Smith, W. M., and P. S. Ghang. 1996. A low power medium access control protocol for portable multi-media systems. *Third International Workshop Mobile Multimedia Communications*, Sept. 25–27.

Tang, Z., and J. J. Garcia-Luna-Aceves. 1999. Hop-reservation multiple access (HRMA) for ad-hoc networks. *IEEE INFOCOM, 1999*.

Toh, C.-K. 2002. *Ad hoc mobile wireless networks: Protocols and systems*. Englewood Cliffs, NJ: Prentice Hall.

Tseng, Y.-C., C.-S. Hsu, and T.-Y. Hsieh. 2002. Power-saving protocols for IEEE 802.11-based multi-hop ad hoc networks. *Proceedings of IEEE INFOCOM, 2002*.

Wattenhofer, R., L. Li, P. Bahl, and Y. M. Wang. 2001. Distributed topology control for power efficient operation in multihop wireless ad hoc networks. *IEEE INFOCOM* 3:1388–1397.

Woesner, H., J. P. Ebert, M. Schlager, and A. Wolisz. 1998. Power saving mechanisms in emerging standards for wireless LANs: The MAC level perspective. *IEEE Personal Communications* 5 (3): 40–48.

Wu, C., and V. O. K. Li. 1987. Receiver-initiated busy-tone multiple access in packet radio networks. *Proceedings of ACM SIGCOMM Conference*, pp. 336–342.

Ye, W., J. Jeidemann, and D. Estrin. 2002. An energy-efficient MAC protocol for wireless sensor networks. *Proceedings of IEEE INFOCOM, 2002*.

3

ROUTING PROTOCOLS

3.1 Introduction

With the advances of wireless communication technology, low-cost and powerful wireless transceivers are widely used in mobile applications. Mobile networks have attracted significant interest in recent years because of their improved flexibility and reduced costs. Compared to wired networks, mobile networks have unique characteristics. In mobile networks, node mobility may cause frequent network topology changes, which are rare in wired networks. In contrast to the stable link capacity of wired networks, wireless link capacity continually varies because of the impacts from transmission power, receiver sensitivity, noise, fading, and interference. Additionally, wireless mobile networks have a high error rate, power restrictions, and bandwidth limitations.

Mobile networks can be classified into infrastructure networks and mobile ad hoc networks according to their dependence on fixed infrastructures. In an infrastructure mobile network, mobile nodes have wired access points (or base stations) within their transmission range. The access points compose the backbone for an infrastructure network. In contrast, mobile ad hoc networks are autonomously self-organized networks without infrastructure support. In a mobile ad hoc network, nodes move arbitrarily; therefore, the network may experience rapid and unpredictable topology changes. Additionally, because nodes in a mobile ad hoc network normally have limited transmission ranges, some nodes cannot communicate directly with each other. Hence, routing paths in mobile ad hoc networks potentially contain multiple hops, and every node in mobile ad hoc networks has the responsibility to act as a router.

Mobile ad hoc networks originated from the US Defense Advanced Research Projects Agency (DARPA) packet radio network (PRNet) and Suran projects. Because they are independent of pre-established infrastructure, mobile ad hoc networks have advantages such as rapid and easy deployment, improved flexibility, and reduced costs. Mobile ad hoc networks are appropriate for mobile applications either in hostile environments, where no infrastructure is available, or in temporarily established mobile applications that are cost crucial. In recent years, application domains of mobile ad hoc networks have gained increasing importance in nonmilitary public organizations and in commercial and industrial areas. The typical application scenarios include rescue missions, law enforcement operations, cooperating industrial robots, traffic management, and educational operations on campus.

Active research work for mobile ad hoc networks is carried on mainly in the fields of medium access control, routing, resource management, power control, and security. Because of the importance of routing protocols in dynamic multihop networks, a lot of mobile ad hoc network routing protocols have been proposed in the last few years. There are some challenges that make the design of mobile ad hoc network routing protocols a tough task:

1. In mobile ad hoc networks, node mobility causes frequent topology changes and network partitions.
2. Because of the variable and unpredictable capacity of wireless links, packet losses may happen frequently.
3. The broadcast nature of the wireless medium introduces the hidden-terminal and exposed-terminal problems.
4. Because mobile nodes have restricted power, computing, and bandwidth resources, ad hoc networks require effective routing schemes.

As a promising network type in future mobile applications, mobile ad hoc networks are attracting more and more researchers. This chapter gives the state-of-the-art review for typical routing protocols for mobile ad hoc networks, including unicast and classical mobile ad hoc network (MANET) unicast and multicast routing algorithms, and popular classification methods.

3.2 Design Issues of Routing Protocols for Ad Hoc Networks

The major design issues in mobile ad hoc networks are discussed in this section.

3.2.1 Routing Architecture

The routing architecture of self-organized networks can be either hierarchical or flat. In most self-organized networks, the hosts will be acting as independent routers, which implies that routing architecture should conceptually be flat (i.e., each address serves only as an identifier and does not convey any information about where one host is topologically located with respect to any other node). In a flat self-organized network, mobility management is not necessary since all of the nodes are visible to each other via routing protocols. In flat routing algorithms such as the destination sequence distance vector (DSDV) and the wireless routing protocol (WRP), the routing tables have entries to all hosts in the self-organized network.

In a flat routing algorithm, routing overhead increases at a faster rate when the size of the network increases. Hence, to control channel reuse spatially (in terms of frequency, time, or spreading code) and reduce routing information overhead, some form of hierarchical scheme should be employed. Clustering is the most common technique employed in hierarchical routing architectures. The idea behind hierarchical routing is to divide the hosts of self-organized networks into a number of overlapping or disjointed clusters. One node is elected as cluster head for each cluster. This cluster head maintains the membership information for the cluster.

Other nodes that are present in the cluster will be treated as ordinary nodes. When these nodes want to send a packet, the nodes can send the packet to the cluster head that routes the packet toward the destination. Cluster head gateway switch routing (CGSR) and the cluster-based routing protocol (CBRP) belong to this type of routing scheme. Hierarchical routing involves cluster management and address/mobility management.

3.2.2 Unidirectional Links Support

Even though it is assumed that every routing protocol is bidirectional, a number of factors will make wireless links unidirectional:

- *Different radio capabilities:* Different nodes can have different transmit powers and receiving power within a network.
- *Interference:* This is due to either hostile jammers or friendly interference, which will reduce a nearby receiver's sensitivity. For example, host X can receive packets from host Y as there is very little interference in X's vicinity. However, Y may be in the vicinity of an interference node and therefore cannot receive packets from X. So, the link between X and Y is directed from Y to X.
- *Message broadcast requirement:* For upward links, satellite-based transmitters are used. The upward links use different types of alternative paths.
- *Mute mode:* An extreme instance, applicable only in tactical mobile networks, is when hosts cannot transmit due to an impending threat. In such a case, they still need to receive information, but cannot participate in bidirectional communications.
- *The state of link direction is time varying:* In a state diagram, if the wireless communication is represented, the state of the wireless link may be either a persistent or a transient phenomenon. The duration of the stay in a particular state may depend on a function of offered traffic, terrain, and energy availability in the nodes.

3.2.3 Usage of Superhosts

It is true that in all the available routing protocols, all the nodes in a particular network will share same bandwidth available for whole network and other facilities. But in some cases, some hosts will include preponderant bandwidth, guaranteed power supply, and high-speed wireless links. Such hosts are referred to as superhosts. For example, a company in a military environment consists of a number of walking soldiers equipped with low-capacity man-pack radios and a few tanks having high-capacity vehicular radios. These types of self-organized networks have two-tier network architectures: backbone area and subarea. The backbone area is composed of superhosts.

It can be assumed that superhosts do not have much mobility compared to normal hosts, because they have to maintain the stability of the backbone. Normal hosts need not make routing decisions. For example, a satellite host (a superhost) can easily collect the routing information from the normal hosts' geographical locations, build the routing table, and propagate these routes. The example is analogous to a person on stage likely having a much better view of the wireless network throughout an auditorium.

3.2.4 Quality of Service (QoS) Routing

For most of the previous routing protocols, optimization was done only on the basis of the metric: hop distance. In the case of datagram service, this parameter may be sufficient. But in the case of MANETs, which are self-organized networks and are dynamic, it may be difficult to perform efficient resource utilization or to execute critical real-time applications in such environments. For these reasons, it is necessary to provide QoS routing support in order to control the total traffic that can flow into the network effectively. In QoS routing, routing will be established between nodes according to resource availability in the network as well as the QoS requirement of flows. For all the requests, they may not have the same QOS parameters. Therefore, QoS routing means that the path selection will be based on availability of resources and efficient resource utilization. Thus, QoS routing will consider multiple constraints and provide better load balance by allocating traffic on different paths, subject to the QoS requirement of different traffic.

On the other hand, current routing protocols seem to favor routing traffic based on shortest path, thereby causing a bottleneck. In self-organized networks, there are many metrics to be considered: (1) most reliable path, (2) most stable path, (3) maximum total power remained path, (4) maximum available bandwidth path, etc. It is desirable to select the routes with minimum cost based on these metrics rather than only to provide the shortest path based on the hop distance.

3.2.5 Multicast Support

In multicast routing, which is a network layer function, data packets from a source reach a group of many destinations. As we know, multicast routing is a network layer function that constructs paths along which data packets from a source are distributed to reach many, but not all, destinations in a communication network. Then, multicast routing sends a single copy of a data packet simultaneously to multiple receivers over a communication link that is shared by the paths to the receivers.

Multicast supports group communication, especially in the case of MANETs, where the network is self-organized, where bandwidth is limited, and where energy is constrained. MANETs consists of several cooperative work groups. The deployment of multicast routing in self-organized networks will provide collaborative visualization and multimedia conference as well as information dissemination in critical situations such as disaster or military scenarios. Multicast routing in self-organized networks became an active research topic only very recently; much research has focused on designing the unicast routing protocols. However, a self-organized network is better suited to multicast than unicast routing because of its broadcast characteristics.

Having multicast routing in self-organized networks poses new challenges. Traditional multicast protocols are not suitable for this environment for the following reasons:

1. The source oriented protocols are inefficient as the source originates the route request moves.
2. As the nodes in the self-organized networks move, they change the topology of the networks; because of this, routing may be difficult.
3. Transient loops may form during spanning tree reconfiguration.
4. Since the communication is to a group of nodes, maintaining too much multicast-related state information puts much pressure on both storage capacity and power, and these resources are severely limited in handheld devices in self-organized networks.

3.3 Classification of Routing Protocols

Designing an efficient and reliable multicast routing protocols is a very challenging problem, because of the MANET characteristics like limited resources. An intelligent routing strategy is required to use limited resources efficiently while at the same time being adaptable to changing network conditions such as network size, traffic density, and network partitioning. Apart from this, it should provide different levels of QoS to different types of applications and users.

In wired networks, link-state and distance-vector algorithms are commonly used. In link-state routing, each node maintains an up-to-date view of the network by periodically broadcasting the link-state costs of its neighboring nodes to all other nodes using a flooding strategy. When each node receives an update packet, it updates its view of the network and its link-state information by applying a shortest-path algorithm to choose the next hop node for each destination.

But these protocols are not well suited for large MANETs. Since the periodic or frequent route updates in large networks may consume a significant part of the available bandwidth, increase channel contention, and may require each node to recharge its power supply frequently. To overcome the problems associated with the protocols of wired networks, a number of routing protocols have been proposed for MANETs. These protocols can be classified into three different groups: global, or proactive; on demand, or reactive; and hybrid.

In proactive routing protocols, the routes to all the destinations (or parts of the network) are determined at the start-up and maintained by using a periodic route update process. In reactive protocols, routes are determined when they are required by the source using a route discovery process. Hybrid routing protocols combine the basic properties of two classes of protocols into one. That is, they are both reactive and proactive in nature. Each group has a number of different routing strategies, which employ a flat or a hierarchical routing structure.

Classification methods are required to help researchers and designers to study, compare and analyze mobile ad hoc routing protocols. These characteristics mainly are related to the information exploited for routing, when this information is acquired, and the roles that nodes may take in the routing process.

3.3.1 Proactive, Reactive, and Hybrid Routing

One of the most popular methods to distinguish mobile ad hoc network routing protocols is based on how routing information is acquired and maintained by mobile nodes. Using this method, mobile ad hoc network routing protocols can be divided into proactive routing, reactive routing, and hybrid routing.

In a proactive routing protocol, nodes in the network calculate routes to all reachable nodes a priori and try to maintain consistent, up-to-date routing information. A proactive routing protocol is also called a "table-driven" protocol. Therefore, a source node can get a routing path immediately if it needs one.

In proactive routing protocols, all nodes have to maintain the information about the network topology. For any change that occurs in the network topology, the updates must be propagated throughout the network to communicate the change. Most proactive routing protocols proposed for mobile ad hoc networks have inherited most of the properties from algorithms used in wired networks. To adapt to the dynamic features of mobile ad hoc networks, necessary modifications have been made on traditional wired network routing protocols.

The major overhead of proactive routing algorithms is whether the request is there or not; regardless, up-to-date network topology is maintained. In the next section, we introduce several typical proactive mobile ad hoc network routing protocols, such as the wireless routing protocol (WRP), the DSDV protocol, and the fisheye state routing (FSR) protocol.

Reactive routing protocols for mobile ad hoc networks are also called "on-demand" routing protocols. In a reactive routing protocol, routing paths are searched only when necessary. A route discovery operation invokes a route-determination procedure. The discovery procedure terminates when either a route has been found or no route is available after examination for all route permutations.

In reactive routing protocols, less control overhead will be there as the routes will not be calculated a priori. Reactive protocols have better scalability than proactive routing protocols as the route calculation is done when the request is made.

In a mobile ad hoc network, active routes may be disconnected due to node mobility. Therefore, route maintenance is an important operation of reactive routing protocols. However, when using reactive routing protocols, source nodes may suffer from long delays for route searching before they can forward data packets. The Dynamic source routing (DSR) and ad hoc on-demand distance vector routing (AODV) are examples for reactive routing protocols for mobile ad hoc networks.

In hybrid routing protocols, the merits of both proactive and reactive routing protocols are combined. In hybrid routing protocols for mobile ad hoc networks, proactive routing approaches are exploited in hierarchical network architectures and reactive routing approaches are exploited in different hierarchical levels. In this chapter, the zone routing protocol (ZRP), zone-based hierarchical link-state (ZHLS) routing protocol, and hybrid ad hoc routing protocol (HARP) will be introduced and analyzed as examples of hybrid routing protocols for mobile ad hoc networks.

3.3.2 Structuring and Delegating the Routing Task

Another classification method is based on the roles that nodes may have in a routing scheme. In a uniform routing protocol, all mobile nodes have the same role, importance, and functionality. Examples of uniform routing protocols include WRP, DSR, AODV, and DSDV. Uniform routing protocols normally assume a flat network structure.

In a nonuniform routing protocol for mobile ad hoc networks, some nodes carry out distinct management and/or routing functions. Normally, distributed algorithms are exploited to select these special nodes. In these routing protocols, routing approaches are related to hierarchical network structures to facilitate node organization and management.

These protocols can be further divided based on the way the organization of the nodes is done and management and routing functions are performed. Following these criteria, nonuniform routing protocols for mobile ad hoc networks are divided into zone-based hierarchical routing, cluster-based hierarchical routing, and core-node-based routing.

In zone-based routing protocols, different zone constructing algorithms are exploited for node organization (e.g., some zone constructing algorithms use geographical information). Dividing the network into zones effectively reduces the overhead for routing information maintenance. Mobile nodes in the same zone know how to reach each other with smaller cost compared to maintaining routing information for all nodes in the whole network. In some zone-based routing protocols, specific nodes act as gateway nodes and carry out interzone communication. The ZRP and ZHLS are zone-based hierarchical routing protocols for mobile ad hoc networks.

A cluster-based routing protocol uses specific clustering algorithms for cluster head election. Mobile nodes are grouped into clusters and cluster heads take the responsibility for membership management and routing functions. CGSR will be introduced in a future section as an example of cluster-based mobile ad hoc network routing protocols. Some cluster-based mobile ad hoc network routing protocols potentially support a multilevel cluster structure, such as hierarchical state routing (HSR).

In core-node-based routing protocols, critical nodes are selected dynamically and carry out special functions, such as routing path construction and control or data packet propagation. Core-extraction distributed ad hoc routing (CEDAR) is a typical core-node-based mobile ad hoc network routing protocol.

3.3.3 Exploiting Network Metrics for Routing

Most of the routing protocols in MANET's use "hop number" as a metric for classifying the routing protocols. This means that if multiple routes are available for the same path, then the path with lowest hop number will be considered.

If all wireless links in the network have the same failure probability, short routing paths are more stable than the long ones and can obviously decrease traffic overhead and reduce packet collisions.

Different mobile applications have different QOS requirements for different characteristics like packet routing and forwarding. QOS routing protocols can use metrics that are used in wired networks, such as bandwidth, delay, delay jitter, packet loss rate, etc. As an example, bandwidth and link stability are used in CEDAR as metrics for routing path construction.

3.3.4 Evaluating Topology, Destination, and Location for Routing

In a topology-based routing protocol for mobile ad hoc networks, nodes collect network topology information for making routing decisions. Other than topology-based routing protocols, there are some destination-based routing protocols proposed in mobile ad hoc networks.

In a destination-based routing protocol, a node only needs to know the next hop along the routing path when forwarding a packet to the destination. For example, DSR is a topology-based routing protocol. AODV and DSDV are destination-based routing protocols.

The availability of global positioning system (GPS) or similar locating systems allows mobile nodes to access geographical information easily. In location-based routing protocols, the distance between a packet forwarding node and the destination, along with the node mobility, can be used in both route discovery and packet forwarding. Existing location-based routing approaches for mobile ad hoc networks can be divided into two schemes. In the first case, the nodes send packets to the destination based on the corresponding node's location information and they will not use any extra information. In the second case, the protocols use both location information and topology information. Location aided routing (LAR) and the distance routing effect algorithm for mobility (DREAM) are typical location-based routing protocols proposed for mobile ad hoc networks.

3.4 Proactive Routing Protocols

A proactive routing protocol is also called a table-driven routing protocol. Using a proactive routing protocol, nodes in a mobile ad hoc network continuously evaluate routes to all reachable nodes and attempt to maintain consistent, up-to-date routing information. Therefore, a source node can get a routing path immediately if it needs one.

In proactive routing protocols, each and every node needs to maintain the up-to-date information about the network topology so that if any link fails or any topology change happens, the information can be propagated to related nodes. Most of the proactive routing protocols discussed for mobile ad hoc networks have inherited properties from algorithms used in wired networks. To adapt to the dynamic features of mobile ad hoc networks, necessary modifications have been made on traditional wired network routing protocols.

Using proactive routing algorithms, mobile nodes proactively update network state and maintain a route regardless of whether data traffic exists or not. The overhead of these types of routing protocols is always knowing the up-to date information about the network. In the following sections, we will discuss several typical proactive mobile ad hoc network routing protocols, such as

- Wireless routing protocol (WRP)
- Destination sequence distance vector (DSDV)
- Fisheye state routing (FSR)
- Hierarchical state routing (HSR)
- Topology broadcast reverse forwarding (TBRF)

3.4.1 Wireless Routing Protocol (WRP)

WRP is a proactive routing protocol that uses the updated version of Bellman–Ford distance vector routing algorithm adaptable to the mobile and ad hoc feature of MANETs. In WRP, each node maintains a distance table and a routing table.

Using WRP, each mobile node maintains a distance table, a routing table, a link-cost table, and a message retransmission list (MRL). An entry in the routing table contains the distance to a destination node, the predecessor, and the successor along the paths to the destination, and a tag to identify its state (i.e., whether it is a simple path, a loop, or invalid). Storing predecessor and successor in the routing table helps to detect routing loops and avoid the counting-to-infinity problem, which is the main shortcoming of the original distance-vector routing algorithm. A mobile node creates an entry for each neighbor in its link-cost table. The entry contains cost of the link connecting to the neighbor and the number of time-outs since an error-free message was received from that neighbor.

In WRP, using update messages, mobile nodes exchange routing. Updated messages can be sent either periodically or whenever link-state changes happen. The MRL contains information about which neighbor has not acknowledged an update message. If needed, the update message will be retransmitted to the neighbor. To ensure connectivity, if there has been no change in its routing table since the last update, a node is required to send a "hello" message. On receiving an update message, the node modifies its distance table and looks for better routing paths according to the updated information.

In WRP, a node checks the consistency of its neighbors after detecting any link change. A consistency check helps to eliminate loops and speed up convergence. If too many tables are to be maintained in WRP, it needs more memory; hence it can be treated as major drawback of this protocol. Moreover, as a proactive routing protocol, it has a limited scalability and is not suitable for large mobile ad hoc networks.

3.4.1.1 Overview To describe the working of the protocol, a network can be modeled as a graph, G (V, E), where V is the set of nodes and E is the set of links connecting the nodes. Each node represents a router and involves components such as a processor, local memory, and input and output queues with unlimited capacity. In a wireless network, a node has radio connectivity with multiple nodes and a single physical radio link connects a node with many other nodes. However, for the purposes of routing-table updating, a node, A, can consider another node, B, as an adjacent node if there is radio connectivity between A and B and A receives update messages from B. Accordingly, we map a physical broadcast link connecting multiple nodes into multiple point-to-point functional links defined for these node paths that are considered to be neighbors of each other. A positive weight is assigned in each direction for a bidirectional link. All messages received (transmitted) by a node are put in an input (output) queue and are processed in first in–first out (FIFO) order. All the update messages are received in the order in which they are transmitted.

WRP is designed so as to work on top of the MAC environment. Updated messages can be lost or sometimes become corrupted because of changes in radio connectivity or jamming. Reliable transmission of update messages is implemented by means of retransmissions. After receiving an update message free of errors, a node is required to send a positive acknowledgment (ACK) indicating that it has good radio connectivity and has processed the update message.

Instead of sending the update message to each and every node, n, of the radio channel, a node can send a single update message to inform all its neighbors about changes in its routing table; however, each such neighbor sends an ACK to the originator node. In addition to ACKs, the connectivity can also be ascertained with the receipt of any message from a neighbor (which need not be an update message). To ensure that connectivity with a neighbor still exists when there are no transmissions or routing table or update ACKs, periodic update messages without any routing table changes are sent to the neighbors. The time interval between two such null update messages is the HelloInterval. If a node fails to receive any type of message from a neighbor for a specified amount of time (e.g., three or four times the HelloInterval, known as the Router-DeadInterval), the node must assume that connectivity with that neighbor has been lost.

3.4.1.2 Information Maintained at Each Node   For the purpose of routing, each node maintains a distance table, a routing table, a link-cost table, and a message retransmission list. The distance table of the node, i, is a matrix containing, for each destination, j, and each neighbor of i (say, k), the distance to j (D^i_{jk}) and the predecessor (p^i_{jk}) reported by k. The routing table of a node i is a vector with an entry for each known destination j that specifies the following:

- The destination's identifier
- The distance to the destination (D^i_j)
- The predecessor of the chosen shortest path to j (p^i_j)
- The successor (s^i_j) of the chosen shortest path to j
- A marker (tag^i_j) used to update the routing table that specifies whether the entry corresponds to a simple path (tag^i_j = correct), a loop (tag^i_j = error), or a destination that has not been marked (tag^i_j = null)

The link-cost table of node i lists the cost of relaying information through each neighbor, k, and the number of periodic update periods that have elapsed since node i received any error-free messages from k. The cost of a failed link is considered to be infinity.

The cost of a link could simply be 1, reflecting the hop count, or the addition of the latency over the link plus some constant bias. The cost of the link from i to k, (i, k), is denoted by l^i_k.

The information about the retransmissions is also maintained in the message retransmission list (MRL), where the *m*th entry consists of the following:

- The sequence number of an update message
- A retransmission counter that is decremented every time node i sends a new update message

- An ACK-required flag (denoted by a^i_{km}) that specifies whether node k has sent an ACK to the update message represented by the retransmission entry
- The list of updates sent in the update message

The preceding information permits node i to know which updates of an update message (each update message contains a list of updates) have to be retransmitted and which neighbors should be requested to acknowledge such retransmission.

Node i retransmits the list of updates in an update message when the retransmission counter of the corresponding entry in the MRL reaches zero. The retransmission counter of a new entry in the MRL is set equal to a small number (e.g., three or four).

3.4.1.3 Information Exchanged among Nodes In WRP, nodes exchange routing-table update messages that propagate only from a node to its neighbors. An update message contains the following information:

- The identifier of the sending node
- A sequence number assigned by the sending node
- An update list of zero or more updates or ACKs to update messages (An update entry specifies a destination, a distance to the destination, and a predecessor to the destination. An ACK entry specifies the source and sequence number of the update message being acknowledged.)
- A response list of zero or more nodes that should send an ACK to the update message

When the update messages are sent, if the space is not large enough to contain all the updates and ACKs that a node wants to report, they are sent in multiple update messages. An example of this event can be the case in which a node identifies a new neighbor and sends its entire routing table.

The response list of the update message is used to avoid the situation in which a neighbor is asked to send multiple ACKs to the same update message, simply because some other neighbor of the node sending the update did not acknowledge.

The first transmission of an update message must ask all neighbors to send an ACK, of course, and this is accomplished by specifying the neighbor's address, which consists the all-neighbors address, which consists of all ones. When the update message reports no updates, the empty address is specified; this address consists of all zeros and instructs the receiving nodes not to send an ACK in return. This type of update message is used as a "hello message" from a node to allow its neighbors to know that they maintain connectivity, even if no user messages or routing-table updates are exchanged.

3.4.1.4 Routing-Table Updating   A node can decide to update its routing table after either receiving an update message from a neighbor or detecting a change in the status of a link to a neighbor. When a node, i, receives an update message from its neighbor, k, it processes each update and ACK entry of the update message in order. In WRP, a node checks the consistency of predecessor information reported by all its neighbors each time it processes an event involving a neighbor, k. In contrast, all previous path-finding algorithms check the consistency of the predecessor only for the neighbor associated with the input event. This unique feature of WRP accounts for its fast convergence after a single resource failure or recovery as it eliminates more temporary looping situations than previous path-finding algorithms.

3.4.2 Destination-Sequence Distance Vector (DSDV)

DSDV is a proactive unicast mobile ad hoc network routing protocol. The DSDV protocol makes us of the traditional Bellman–Ford algorithm. However, in order to adopt to mobile ad hoc networks, the routing mechanisms are improved. In routing tables of DSDV, an entry stores the next hop toward a destination, the cost metric for the routing path to the destination, and a destination sequence number that is created by the destination. To distinguish stale paths from fresh ones, sequence numbers are used; sequence numbers also avoid route loops.

DSDV can have routing updates based on either time or event. Every node periodically transmits updates including its routing information to its immediate neighbors. A routing table can be updated in two ways: a "full dump" (the full routing table is included inside the update) or an incremental update (which contains only those entries that, with a metric, have been changed since the last update was sent).

Routing information is advertised by broadcasting or multicasting the packets, which are transmitted periodically and incrementally as topological changes are detected (e.g., when stations move within the network). Data are also kept about the length of time between arrival of the first and arrival of the best route for each particular destination. Based on these data, a decision may be made to delay advertising routes that are about to change soon, thus damping fluctuations of the route tables. The advertisement of routes that may not have stabilized yet is delayed in order to reduce the number of rebroadcasts of possible route entries that normally arrive with the same sequence number.

The DSDV protocol requires each mobile station to advertise its own routing table to each of its current neighbors (for instance, by broadcasting its entries). The entries in this list may change fairly dynamically over time, so the advertisement must be made often enough to ensure that every mobile computer can almost always locate every other mobile computer of the collection.

In addition, each mobile computer agrees to relay data packets to other computers upon request. This agreement places a premium on the ability to determine the shortest number of hops for a route to a destination; if the nodes are in sleep node, they are not disturbed. In this way a mobile computer may exchange data with any other mobile computer in the group, even if the target of the data is not within range for direct communication.

If the notification of which other mobile computers are accessible from any particular computer in the collection is done at layer 2, then DSDV will work with whatever higher layer (e.g., network layer) protocol might be in use.

The data broadcast by each mobile computer will contain its new sequence number and the following information for each new route:

- The destination's address
- The number of hops required to reach the destination
- The sequence number of the information received regarding that destination, as originally stamped by the destination

The routing table also contains the network address and the hardware address of the node transmitting the routing tables, within the headers of the packet. The routing table will also include a sequence number created by the transmitter. Routes with more recent sequence numbers will be preferred for making the forwarding decisions.

A route with a sequence number equal to an existing route is chosen if it has a "better" metric and the existing route is discarded or stored as less preferable. The metrics for routes chosen from the newly received broadcast information are each incremented by one hop. Newly recorded routes are scheduled for immediate advertisement to the current mobile host"s neighbors. Routes that show an improved metric are scheduled for advertisement at a time that depends on the average settling time for routes to the particular destination under consideration.

Advantages of DSDV include the following:

- DSDV is an efficient protocol for route discovery.
- Route discovery latency is very low.
- Loop-free paths are guaranteed in DSDV.

Disadvantages of DSDV include:

- To maintain network topology at each node, DSDV needs to send a lot of control messages.
- DSDV generates a high volume of traffic for high-density and highly mobile networks.

3.4.3 Fisheye State Routing (FSR)

FSR is an implicit hierarchical routing protocol. It uses the "fisheye" technique proposed by Kleinrock and Stevens. In this technique, they tried to reduce the size of information required to represent graphical data. The eye of a fish captures with high detail the pixels near the focal point. The detail decreases as the distance from the focal point increases. In routing, the fisheye approach translates to maintaining accurate distance and path quality information about the immediate neighborhood of a node, with progressively less detail as the distance increases.

The functionality of FSR is similar to linked state routing (LSR), where it maintains a topology map at each node. The difference with LSR is the way in which routing information is disseminated. In LSR, whenever a topology change happens, link-state packets are generated and flooded into the network. But in case of FSR, link-state packets are not flooded. Instead, maintain a link-state table is maintained based on the up-to-date information received from neighboring nodes, and they periodically exchange it with their local neighbors only (no flooding). Because of this exchange process, the table entries with larger sequence numbers replace the ones with smaller sequence numbers.

The method of exchange used in FSR resembles the vector exchange in distributed Bellman–Ford (DBF) (or, more precisely, DSDV), where the distance updates happen through the time stamp or sequence number assigned by the node originating the update. But in the case of FSR, link states rather than distance vectors are propagated. Moreover, as in LSR, a full topology map is kept at each node and shortest paths are computed using this map.

In the case of a wireless environment, a radio link between mobile nodes may experience frequent disconnects and reconnects. The LSR protocol releases a link-state update for each such change, which floods the network and causes excessive overhead. FSR avoids this problem by using periodic, instead of event-driven, exchange of the topology map, which reduces the control message overhead. When network size grows large, the update message could consume a considerable amount of bandwidth, which depends on the update period. Because of this, in order to reduce the size of update messages without seriously affecting routing accuracy, FSR uses the fisheye technique. Figure 3.1 illustrates the application of fisheye in a mobile wireless network.

In this figure, the circles with different shades of gray define the fisheye scopes with respect to the center node (node 11). The scope is defined as the set of nodes that can be reached within a given number of hops. In our case, three scopes are shown for one, two, and more than two hops, respectively. Nodes are color coded as black, gray, and white accordingly. The number of levels and the radius of each scope will depend on the size of the network. The reduction of routing overhead is obtained by using different exchange periods for different entries in the routing table.

More precisely, entries corresponding to nodes within the smaller scope are propagated to the neighbors with the highest frequency. Referring to Figure 3.2, entries in bold are exchanged most frequently. The rest of the entries are sent out at a lower frequency. As a

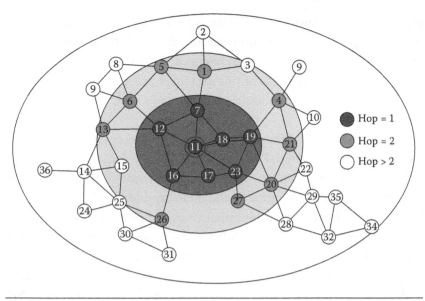

Figure 3.1 Scope of fisheye.

result, a considerable fraction of link-state entries are suppressed in a typical update, thus reducing the message size. This strategy produces timely updates from near stations, but creates large latencies from stations afar. However, the imprecise knowledge of the best path to a distant destination is compensated by the fact that the route becomes progressively more accurate as the packet gets closer to its destination. As the network size grows large, a "graded" frequency update plan must be used across multiple scopes to keep the overhead low.

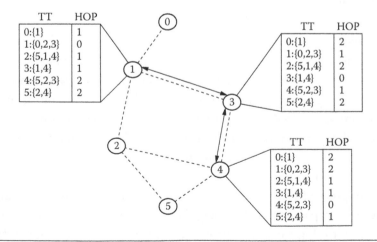

Figure 3.2 Message reduction using fisheye.

The FSR concept originates from the global state routing (GSR) protocol. GSR can be viewed as a special case of FSR in which there is only one fisheye scope level. As a result, the entire topology table is exchanged among neighbors. Clearly, this consumes a considerable amount of bandwidth when network size becomes large. Through updating link-state information with different frequencies, depending on the scope distance, FSR scales well to large network size and keeps overhead low without compromising route computation accuracy when the destination is near.

By retaining a routing entry for each destination, FSR avoids the extra work of "finding" the destination (as in on-demand routing) and thus maintains low single-packet transmission latency. As mobility increases, routes to remote destinations become less accurate. However, when a packet approaches its destination, it finds increasingly accurate routing instructions as it enters sectors with a higher refresh rate.

3.4.4 Ad Hoc On-Demand Distance Vector (AODV)

An ad hoc network is a distributed system without any centralized access point or framed infrastructure. AODV is an algorithm working for the operation of such ad hoc networks. Each mobile host operates as a specialized router and routes are obtained as needed. The AODV routing algorithm is quite suitable for a dynamic self-starting network as required by users wishing to utilize ad hoc networks. AODV provides loop-free routes even while repairing broken links. Because the protocol does not require global periodic routing advertisements, the demand is less than in those protocols that do necessitate such advertisements.

3.4.4.1 Path Discovery The path discovery process is initiated whenever a source node needs to communicate with another node for which it has no routing information in its table. Every node maintains two separate counters: a node sequence number and a broadcast ID. The source node initiates path discovery by broadcasting a route request (RREQ) packet to its neighbors. The RREQ contains the following fields:

<source_addr source sequence# broadcast id dest_addr dest sequence# hop cnt >

The pair *<source_addr, broadcast_id>* uniquely identifies an RREQ. *broadcast_id* is incremented whenever the source issues a new RREQ. Each neighbor either satisfies the RREQ by sending a route reply (RREP) back to the source or broadcasts the RREQ to its own neighbors after increasing the *hop_cnt*. Notice that a node may receive multiple copies of the same route broadcast packet from various neighbors. When an intermediate node receives an RREQ, if it has already received an RREQ with the same *broadcast_id* and source address, it drops the redundant RREQ and does not rebroadcast it.

If a node cannot satisfy the RREQ, it keeps track of the following information in order to implement the reverse path setup as well as the forward path setup that will accompany the transmission of the eventual RREP:

<dest_ IP_addr, Source_IP_addr, Broadcast_ID, Expiration_ time, Source_sequence#>

3.4.4.2 Reverse Path Setup   There are two sequence numbers (in addition to the broadcast_id) included in an RREQ: the source sequence number and the last destination sequence number known to the source. The source sequence number is used to maintain freshness information about the reverse route to the source; the destination sequence number specifies how fresh a route to the destination must be before it can be accepted by the source. As the RREQ travels from a source to various destinations, it automatically sets up the reverse path from all nodes back to the source, as illustrated in Figure 3.3. To set up a reverse path, a node records the address of the neighbor from which it received the first copy of the RREQ. These reverse path route entries are maintained for at least enough time for the RREQ to traverse the network and produce a reply to the sender.

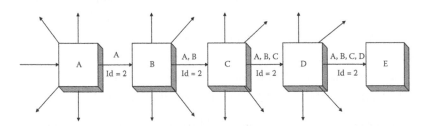

Figure 3.3   Route discovery example: node A is the initiator, and node E is the target.

3.4.4.3 Forward Path Setup   Eventually, an RREQ will arrive at a node (possibly the destination itself) that possesses a current route to the destination. The receiving node first checks that the RREQ was received over a bidirectional link. If an intermediate node has a route entry for the desired destination, it determines whether the route is current by comparing the destination sequence number in its own route entry to the destination sequence number in the RREQ. If the RREQ's sequence number for the destination is greater than that recorded by the intermediate node, the intermediate node must not use its recorded route to respond to the RREQ. Instead, the intermediate node rebroadcasts the RREQ.

The intermediate node can reply only when it has a route with a sequence number that is greater than or equal to that contained in the RREQ. If it does have a current route to the destination and if the RREQ has not been processed previously, the node then unicasts an RREP packet back to the neighbor from which it received the RREQ. An RREP contains the following information:

<source_addr, dest_addr, dest_sequence #, hop_cnt, lifetime>

By the time a broadcast packet arrives at a node that can supply a route to the destination, a reverse path has been established to the source of the RREQ. As the RREP travels back to the source, each node along the path sets up a forward pointer to the node from which the RREP came, updates its time-out information for route entries to the source and destination, and records the latest destination sequence number for the requested destination. Figure 3.4 represents the forward path setup as the RREP travels from the destination, D, to the source node, S. Nodes that are not along the path determined by the RREP will time out after ACTIVE_ROUTE_TIMEOUT(3000 ms) and will delete the reverse pointers.

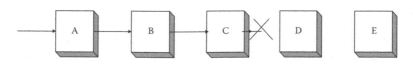

Figure 3.4   Route maintenance example: Node C is unable to forward a packet from A to E over its link to next-hop D.

A node receiving an RREP propagates the first RREP for a given source node toward that source. If it receives further RREPs, it updates its routing information and propagates the RREP only if the RREP contains either a greater destination sequence number than the previous RREP or the same destination sequence number with a smaller hop count. It suppresses all other RREPs received. This decreases the number of RREPs propagating toward the source while also ensuring the most up-to-date and quickest routing information. The source node can begin data transmission as soon as the first RREP is received and can later update its routing information if it learns of a better route.

3.4.4.4 Route Table Management   In addition to the source and destination sequence numbers, other useful information is also stored in the route table entries and is called the soft state associated with the entry. Associated with reverse path routing entries is a timer, called the route request expiration timer. The purpose of this timer is to purge reverse path routing entries from those nodes that do not lie on the path from the source to the destination. The expiration time depends upon the size of the ad hoc network.

Another important parameter associated with routing entries is the route caching time-out, or the time after which the route is considered to be invalid. In each routing table entry, the address ofactive neighbors through which packets for the given destination are received is also maintained. A neighbor is considered active for that destination if it originates or relays at least one packet for that destination within the most recent active time-out period. This information is maintained so that all active source nodes can be notified when a link along a path to the destination breaks. A route entry is considered active if it is in use by any active neighbors. The path from a source to a destination, which is followed by packets along active route entries, is called an "active path." A mobile node maintains a route table entry for each destination of interest. Each route table entry contains the following information:

Another important parameter associated with routing entries is the route caching time-out, or the time after which the route is considered to be invalid. In each routing table entry, the address of active neighbors through which packets for the given destination are received is also maintained. A neighbor is considered active for that destination if it originates or relays at least one packet for that destination within the most recent active time-out period. This information is maintained so that all active source nodes can be notified when a link along a path to the destination breaks. A route entry is considered active if it is in use by any active neighbors. The path from a source to a destination, which is followed by packets along active route entries, is called an "active path." A mobile node maintains a route table entry for each destination of interest. Each route table entry contains the following information:

- Destination
- Next hop
- Number of hops (metric)
- Sequence number for the destination
- Active neighbors for this route
- Expiration time for the route table entry

Each time a route entry is used to transmit data from a source toward a destination, the time-out for the entry is reset to the current time plus active route time-out. If a new route is offered to a mobile node, the mobile node compares the destination sequence number of the new route to the destination sequence number for the current route. The route with the greater sequence number is chosen. If the sequence numbers are the same, then the new route is selected only if it has a smaller metric (fewer numbers of hops) to the destination.

3.4.4.5 Path Maintenance Movement of nodes not lying along an active path does not affect the routing to that path's destination. If the source node moves during an active session, it can reinitiate the route discovery procedure to establish a new route to the destination. When either the destination or some intermediate node moves, a special RREP is sent to the affected source nodes. Periodic hello messages can be used to ensure symmetric links, as well as to detect link failures.

Once the next hop becomes unreachable, the node upstream of the break propagates an unsolicited RREP with a fresh sequence number—that is, a sequence number that is one greater than the previously known sequence number—and hop count to all active upstream neighbors. Those nodes subsequently relay that message to their active neighbors and so on. This process continues until all active source nodes are notified. It terminates because AODV maintains only loop-free routes and there are only a finite number of nodes in the ad hoc network.

When the source node receives the broken link if the source still requires restarting of the discovery process of finding the route to destination, it can start. To determine whether a route is still needed, a node may check whether the route has been used recently as well as inspect upper level protocol control blocks to see whether connections remain open using the indicated destination. If the source node or any other node along the previous route decides that it would like to rebuild the route to the destination, it sends out an RREQ with a destination sequence number of one greater than the previously known sequence number to ensure that it builds a new viable route and that no nodes reply if they still regard the previous route as valid.

3.4.4.6 Local Connectivity Management   Whenever a node receives a broadcast from a neighbor it updates its local connectivity information to ensure that it includes this neighbor. In the event that a node has not sent any packets to all of its active downstream neighbors within a hello interval, it broadcasts a hello message to its neighbors. The node's sequence number is not changed for hello message transmissions. This hello message is prevented from being rebroadcast outside the neighborhood of the node because it contains a Time_To_Live (TTL) value of one. Neighbors that receive this packet update their local connectivity information to the node.

Receiving a broadcast or a hello from a new neighbor or failing to receive allowed hello loss consecutive hello messages from a node previously in the neighborhood is an indication that the local connectivity has changed.

3.4.5 Dynamic Source Routing (DSR) Protocol

The DSR protocol is a simple and efficient routing protocol designed specifically for use in multihop wireless ad hoc networks. In DSR each node dynamically discovers a path to the destination. Each data packet sent carries in its header the complete information about the list of nodes that it has to traverse to reach the destination and avoids the need for up-to-date routing information in the intermediate nodes through which the packet is forwarded. By including this source route in the header of each data packet, other nodes forwarding or over-hearing any of these packets may also easily cache this routing infor-mation for future use.

3.4.5.1 Overview and Important Properties of the Protocol The DSR pro-tocol is composed of two mechanisms that work together to allow the discovery and maintenance of source routes in the ad hoc network:

- *Route discovery* is the mechanism by which a node, S, wishing to send a packet to a destination node, D, obtains a source route to D. Route discovery is used only when S attempts to send a packet to D and does not already know a route to D.
- *Route maintenance* is the mechanism by which node S is able to detect, while using a source route to D, if the network topology has changed such that it can no longer use its route to D because a link along the route no longer works. When route maintenance indicates that a source route is broken, S can attempt to use any other route it happens to know to D, or it can invoke route discovery again to find a new route. Route maintenance is used only when S is actually sending packets to D.

3.4.5.2 Basic DSR Route Discovery When some node S originates a new packet destined to some other node D, it places in the header of the packet a *source route* giving the sequence of hops that the packet should follow on its way to D. Normally, S will obtain a suitable source route by searching its *route cache* of routes previously learned, but if no route is found in its cache, it will initiate the route discovery protocol to find a new route to D dynamically.

In this case, we call S the *initiator* and D the *target* of the route dis-covery. For example, Figure 3.3 illustrates an example route discovery in which a node A is attempting to discover a route to node E. To ini-tiate the route discovery, A transmits a ROUTE REQUEST message as a single local broadcast packet, which is received by (approximately) all nodes currently within wireless transmission range of A. Each ROUTE REQUEST message identifies the initiator and target of the route discovery and contains a unique *request id* determined by the initiator of the REQUEST. Each ROUTE REQUEST also contains a record listing the address of each intermediate node through which this particular copy of the ROUTE REQUEST message has been forwarded. This route record is initialized to an empty list by the initiator of the route discovery.

When another node receives a ROUTE REQUEST, if it is the target of the route discovery, it returns a ROUTE REPLY message to the initiator of the route discovery, giving a copy of the accumulated route record from the ROUTE REQUEST; when the initiator receives this ROUTE REPLY, it caches this route in its route cache for use in sending subsequent packets to this destination.

If this node receiving the ROUTE REQUEST has recently seen another ROUTE REQUEST message from this initiator bearing this same request ID, or if it finds that its own address is already listed in the route record in the ROUTE REQUEST message, it discards the REQUEST. Otherwise, this node appends its own address to the route record in the ROUTE REQUEST message and propagates it by transmitting it as a local broadcast packet (with the same request ID).

In returning the ROUTE REPLY to the initiator of the route discovery, such as node E replying back to A in Figure, node E will typically examine its own route cache for a route back to A, and if one is found, will use it for the source route for delivery of the packet containing the ROUTE REPLY. Otherwise, E may perform its own route discovery for target node A, but to avoid possible infinite recursion of route discoveries, it must piggyback this ROUTE REPLY on its own ROUTE REQUEST message for A. It is also possible to piggyback other small data packets, such as a transmission control protocol (TCP) synchronization (SYN) packet, on a ROUTE REQUEST using this same mechanism. Node E could also simply reverse the sequence of hops in the route record that it is trying to send in the ROUTE REPLY and use this as the source route on the packet carrying the ROUTE REPLY itself.

For MAC protocols such as IEEE 802.11 that require a bidirectional frame exchange as part of the MAC protocol, this route reversal is preferred as it avoids the overhead of a possible second route discovery, and it tests the discovered route to ensure that it is bidirectional before the route discovery initiator begins using the route. However, this technique will prevent the discovery of routes using unidirectional links. In wireless environments where the use of unidirectional links is permitted, such routes may in some cases be more efficient than those with only bidirectional links, or may be the only way to achieve connectivity to the target node.

When a route discovery is initiated, the source node maintains a copy of the original packet in a local buffer called the send buffer. The send buffer contains a copy of each packet that cannot be transmitted by this node because it does not yet have a source route to the packet's destination. Each packet in the send buffer is stamped with the time that it was placed into the buffer and is discarded after residing in the send buffer for some time-out period; if necessary for preventing the send buffer from overflowing, a FIFO or other replacement strategy can also be used to evict packets before they expire.

While a packet remains in the send buffer, the node should occasionally initiate a new route discovery for the packet's destination address. However, the node must limit the rate at which such new route discoveries for the same address are initiated, since it is possible that the destination node is not currently reachable. In particular, due to the limited wireless transmission range and the movement of the nodes in the network, the network may at times become partitioned, meaning that there is currently no sequence of nodes through which a packet could be forwarded to reach the destination. Depending on the movement pattern and the density of nodes in the network, such network partitions may be rare or may be common.

If a new route discovery was initiated for each packet sent by a node in such a situation, a large number of unproductive ROUTE REQUEST packets would be propagated throughout the subset of the ad hoc network reachable from this node.

To reduce the overhead from such route discoveries, we use exponential back-off to limit the rate at which new route discoveries may be initiated by any node for the same target. If the node attempts to send additional data packets to this same node more frequently than this limit, the subsequent packets should be buffered in the send buffer until a ROUTE REPLY is received, but the node must not initiate a new route discovery until the minimum allowable interval between new route discoveries for this target has been reached. This limitation on the maximum rate of route discoveries for the same target is similar to the mechanism required by Internet nodes to limit the rate at which ad hoc routing protocol REQUESTs are sent for any single target IP address.

3.4.5.3 Basic DSR Route Maintenance   When originating or forwarding a packet using a source route, each node transmitting the packet is responsible for confirming that the packet has been received by the next hop along the source route; the packet is retransmitted (up to a maximum number of attempts) until this confirmation of receipt is received. For example, in the situation illustrated in Figure, node A has originated a packet for E using a source route through intermediate nodes B, C, and D. In this case, node A is responsible for receipt of the packet at B, node B is responsible for receipt at C, node C is responsible for receipt at D, and node D is responsible for receipt finally at the destination E.

The confirmation of receipt can be provided as in a standard part of the MAC protocol or by a possible acknowledgment (e.g., node B confirms whether the packet has reached node C by overhearing at the time when C forwards the packet to D). If neither of these confirmation mechanisms is available, the node transmitting the packet may set a bit in the packet's header to request a DSR-specific software acknowledgment be returned by the next hop. This software acknowledgment will normally be transmitted directly to the sending node, but if the link between these two nodes is unidirectional, this software acknowledgment may travel over a different, multihop path.

If the packet is retransmitted by some hop the maximum number of times and no receipt confirmation is received, this node returns a ROUTE ERROR message to the original sender of the packet, identifying the link over which the packet could not be forwarded.

For example, in Figure 3.3, if C is unable to deliver the packet to the next hop D, then C returns a ROUTE ERROR to A, stating that the link from C to D is currently "broken." Node A then removes this broken link from its cache; any retransmission of the original packet is a function for upper layer protocols such as TCP.

For sending such a retransmission or other packets to this same destination E, if A has another route to E in its route cache (for example, from additional ROUTE REPLYs from its earlier route discovery, or from having overheard sufficient routing information from other packets), it can send the packet using the new route immediately. Otherwise, it may perform a new route discovery for this target.

3.4.6 Temporally Ordered Routing Algorithm (TORA)

TORA is a distributed routing protocol. Its intended use is for routing of Internet protocol (IP) datagrams within an autonomous system. The protocol's reaction is structured as a temporally ordered sequence of diffusing computations, each computation consisting of a sequence of directed link reversals. The protocol is highly adaptive, efficient, and scalable; it is well suited for use in large, dense mobile networks.

Depending upon the topological changes, ordering of the algorithm reaction will change subsequently in this protocol. In order to suit the operation in various environmental challenges, TORA was designed with the following properties:

- Distributed in nature
- Provides loop-free and multiple routes
- Establishes routes quickly
- Minimizes communication

TORA can be separated into three basic functions: creating routes, maintaining routes, and erasing routes:

- Creating a route from a given node to the destination requires establishment of a sequence of directed links leading from the node to the destination. This function is only initiated when a node with no directed links requires a route to the destination. Thus, creating routes essentially corresponds to assigning directions to links in an undirected network or portion of the network. The method used to accomplish this is an adaptation of the query/reply process, which builds a directed acyclic graph (DAG) rooted at the destination.

- Maintaining routes refers to reacting to topological changes in the network in a manner such that routes to the destination are re-established within a finite time—meaning that its directed portions return to a destination-oriented DAG within a finite time.
- This leads to the third function: erasing routes. Upon detection of a network partition, all links must be marked as undirected to erase invalid routes.

TORA accomplishes these three functions through the use of three distinct control packets: query (QRY), update (UPD), and clear (CLR). QRY packets are used for creating routes, UPD packets are used for both creating and maintaining routes, and CLR packets are used for erasing routes.

3.4.7 Cluster-Based Routing Protocol (CBRP)

CBRP is a routing protocol designed for mobile ad hoc networks. The protocol divides the nodes of the ad hoc network into two-hop-diameter clusters in a distributed manner. A cluster head is elected for each cluster to maintain cluster information. Intercluster routes are discovered dynamically using the cluster membership information kept at each cluster head. Because of the cluster concept, this protocol efficiently minimizes the flooding traffic during route discovery and speeds up this process as well. Furthermore, this protocol works for unidirectional links and uses these links for both intracluster and intercluster routing.

The two major new features that have been added to the protocol are route shortening and local repair. Both features make use of the two-hop topology information maintained by each node through the broadcasting of HELLO messages. The route shortening mechanism dynamically shortens the source route of the data packet being forwarded and informs the source about the better route. Local route repair patches a broken source route automatically and avoids route rediscovery by the source.

When a routing protocol is designed for ad hoc networks, the many challenges of dynamically changing topology and infrastructure that make IP subnetting inefficient. However, routing protocols that are flat (i.e., have no hierarchy) might suffer from excessive overhead when scaled up. Finally, links in mobile networks could be asymmetric at times.

CBRP has the following features:

- It is a fully distributed operation.
- There is less flooding traffic during the dynamic route discovery process.
- There is explicit exploitation of unidirectional links that would otherwise be unused.
- Broken routes can be repaired locally without rediscovery.
- Suboptimal routes can be shortened as they are used.

In these protocols, clusters are introduced to minimize updating overhead during topology change. However, the overhead for maintaining up-to-date information about the whole network's cluster membership and intercluster routing information at each and every node in order to route a packet is considerable. As network topology changes from time to time due to node movement, the effort to maintain such up-to-date information is expensive and rarely justified, as such global cluster membership information is obsolete long before it is used.

3.4.8 Location-Aided Routing (LAR)

The LAR protocols use location information to reduce the search space for a desired route. Limiting the search space results in fewer route discovery messages.

3.4.8.1 Route Discovery Using Flooding In order to improve performance of routing protocols by using location information, we show how a route discovery protocol based on *flooding* can be improved by describing the route discovery algorithm using flooding.

Consider that a source node, S, needs to find a route to destination node D. Node S broadcasts a *route request* (RREQ) message to all its neighbors. On receiving a route request message, a node (say, X) compares the desired destination with its own identifier. If a comparison matches, then the request is for a route to itself (i.e., node X). Otherwise, node X broadcasts the request to its neighbors

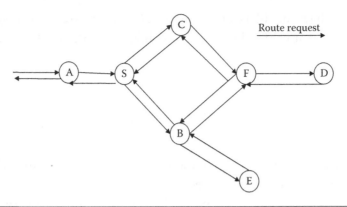

Figure 3.5   Illustration of location aided routing (LAR).

to avoid redundant transmissions of route requests. Node X only broadcasts a particular route request once (repeated reception of a route request is detected using sequence numbers). Figure 3.5 illustrates this algorithm. In this figure, node S needs to determine a route to node D. Therefore, node S broadcasts a route request to its neighbors.

When nodes B and C receive the route request, they forward it to all their neighbors. When node F receives the route request from B, it forwards the request to its neighbors. However, when node F receives the same route request from C, node F simply discards the route request. As the route request is propagated to various nodes, the path followed by the request is included in the route request packet. Using the flooding algorithm, provided that the intended destination is reachable from the sender, the destination should eventually receive a route request message.

On receiving the route request, the destination responds by sending a *route reply* message to the sender; this message follows a path that is obtained by reversing the path followed by the route request received by D (the route request message includes the path traversed by the request). The destination may not receive the route request due to unreachability of a node or due to loss of route request because of transmission errors. In such cases, the sender needs to be able to reinitiate route discovery. Therefore, when a sender initiates route discovery, it sets a time-out. If, within the time-out interval a route reply is not reached, then a new route discovery is initiated. Time-out may occur if the destination does not receive a route request, or if the route reply message from the destination is lost.

3.4.9 Ant-Colony-Based Routing Algorithm (ARA)

ARA is a new on-demand routing algorithm for ad hoc networks. The protocol is based on group intelligence—especially on the ant colony base. This routing protocol is highly adaptive, efficient, and scalable. The main goal in the design of the protocol was to reduce the overhead for routing. Ant algorithms are multiagent systems consisting of agents with the behavior of individual ants, as shown in Figure 3.6.

3.4.9.1 Basic Ant Algorithm   The basic idea of the ant colony optimization is taken from the food searching behavior of ants. When ants are on their way to search for food, they start from their nest and walk toward the food. When an ant reaches an intersection, it has to decide which branch to take next. While walking, ants deposit pheromone, which marks the route taken. The concentration of pheromone deposited on a certain path is an indication of its usage. With time the concentration of pheromone decreases due to diffusion effects.

This property is important because it is integrating a dynamic component into the path-searching process. Figure 3.6 shows a scenario with two routes from the nest to the food place. At the intersection, the first ants randomly select the next branch. Because the lower route is shorter than the upper one, the ants that take this path will reach the food place first. On their way back to the nest, the ants again have to select a path. After a short time, the pheromone concentration on the shorter path will be higher than on the longer path, because the ants using the shorter path will increase the pheromone concentration faster. The shortest path will thus be identified and eventually all ants will use this one.

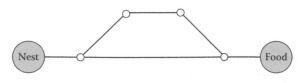

Figure 3.6   Behavior of individual ants.

This behavior of the ants can be used to find the shortest path in networks. Especially, the dynamic component of this method allows a high adaptation to changes in mobile ad hoc network topology, since in these networks the existence of links is not guaranteed and link changes occur very often.

3.5 Hybrid Routing Protocols

Hybrid routing protocols are a new generation of protocol that are both proactive and reactive in nature. These protocols are designed to increase scalability by allowing nodes with close proximity to work together to form some sort of a backbone to reduce the route discovery overheads. In hybrid routing protocols, a proactive method is employed to maintain routes for nearby nodes; a reactive or route discovery method is used for faraway nodes. Most hybrid protocols proposed to date are zone based, which means that the network is partitioned or seen as a number of zones by each node. Others group nodes are formed into trees or clusters. A number of different hybrid routing protocols have been proposed for MANETs:

- Zone routing protocol (ZRP)
- Zone-based hierarchical link state (ZHLS)
- Scalable location updates routing protocol (SLURP)
- Distributed spanning trees-based routing protocol (DST)
- Distributed dynamic routing (DDR)

3.5.1 Zone Routing Protocol (ZRP)

The advantages of both proactive and reactive approaches are combined by maintaining an up-to-date topological map of each zone centered at each node. Within the zone, routes can be immediately found by using a proactive method.

3.5.1.1 Motivation Proactive routing uses excess bandwidth to maintain routing information, while reactive routing involves long route-request delays. Reactive routing also inefficiently floods the entire network for route determination. ZRP aims to address the problems by combining the best properties of both approaches. ZRP can be classed as a hybrid reactive/proactive routing protocol.

In an ad hoc network, it can be assumed that the largest part of the traffic is directed to nearby nodes. Therefore, ZRP reduces the proactive scope to a zone centered on each node. In a limited zone, the maintenance of routing information is easier. Further, the amount of routing information that is never used is minimized. Still, nodes farther away can be reached with reactive routing. Since all nodes proactively store local routing information, route requests can be more efficiently performed without querying all the network nodes.

In spite of using the zones, ZRP makes use of flat view over the network. Because of this, the organization overhead of maintaining hierarchical networks can be avoided. Nodes belonging to different subnets must send their communication to a subnet that is common to both nodes. This may congest parts of the network. ZRP can be categorized as a flat protocol because the zones overlap. Hence, optimal routes can be detected and network congestion can be reduced.

3.5.1.2 Architecture The zone routing protocol, as its name implies, is based on the concept of zones. For each node, a separate routing zone is specified, and the zones of neighboring nodes overlap. The routing zone has a radius ρ expressed in hops. The zone thus includes the nodes, whose distance from the node in question is, at most, h hops. An example routing zone is shown in Figure 3.7, where the routing zone of S includes the nodes A–I, but not K. In the illustration, the radius is marked as a circle around the node in question.

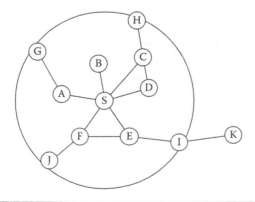

Figure 3.7 Example of a routing zone with r = 2.

The nodes of a zone are divided into peripheral nodes and interior nodes. The nodes whose minimum distance to the central node is exactly equal to the zone radius ρ are called peripheral nodes. The nodes whose minimum distance is less than ρ are interior nodes. In Figure 3.7, the nodes A–F are interior nodes; the nodes G–J are peripheral nodes and the node K is outside the routing zone. Note that node H can be reached by two paths: one with a length two hops and one with a length of three hops. The node is, however, within the zone because the shortest path is less than or equal to the zone radius.

By adjusting the transmission power of the nodes, the number of nodes in the routing zone can be regulated. Lowering the power reduces the number of nodes within direct reach and vice versa. The number of neighboring nodes should be sufficient to provide adequate reachability and redundancy. On the other hand, a too large coverage results in many zone members and the update traffic becomes excessive. Further, large transmission coverage adds to the probability of local contention.

ZRP refers to the locally proactive routing protocol as the intrazone routing protocol (IARP) and the globally reactive routing protocol as the interzone routing protocol (IERP). Instead of broadcasting packets, ZRP uses a concept called *bordercasting*. Bordercasting directs query requests to the border of the zone by the information provided by IARP. The bordercast packet delivery service is provided by the bordercast resolution protocol (BRP). BRP uses a map of an extended routing zone to construct bordercast trees for the query packets. Alternatively, it uses source routing based on the normal routing zone. By employing *query control* mechanisms, route requests can be directed away from areas of the network that already have been covered.

ZRP relies on a neighbor discovery protocol (NDP) to detect new neighbor nodes and link failures provided by the media access control (MAC) layer. NDP transmits "HELLO" beacons at regular intervals. Upon receiving a beacon, the neighbor table is updated. Neighbors for which no beacon has been received within a specified time are removed from the table. If the MAC layer does not include an NDP, the functionality must be provided by IARP. The relationship between the components is illustrated in Figure 3.8. Route updates are triggered by NDP, which notifies IARP when the neighbor table is updated. IERP uses the routing table of IARP to respond to route

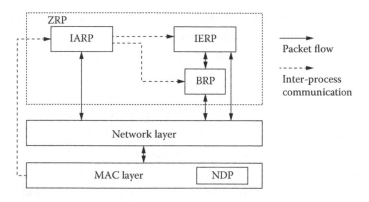

Figure 3.8 Zone routing protocol architecture.

queries. IERP forwards queries with BRP, which uses the routing table of IARP to guide route queries away from the query source.

3.5.1.3 Routing if any node wants to send the packet it first checks whether the destination is within its local zone, using information provided by IARP. If it is within the zone, the packet can be routed proactively. Reactive routing is used if the destination is outside the zone.

The process of reactive routing is divided into two phases: the *route request* phase and the *route reply* phase. In the route request, the source sends a route request packet to its peripheral nodes using BRP. If the receiver of a route request packet knows the destination, it responds by sending a route reply back to the source. Otherwise, it continues the process by bordercasting the packet. In this way, the route request spreads throughout the network.

If several copies of the same route request are received by a node, these are considered redundant and are discarded. The reply is sent by any node that can provide a route to the destination. To be able to send the reply back to the source node, routing information must be collected when the request is sent through the network. The information is recorded either in the route request packet or as next-hop addresses in the nodes along the path.

The zone radius is an important property for the performance of ZRP. If a zone radius of one hop is used, routing is purely reactive. If the radius approaches infinity, routing is reactive. The selection of radius is a trade-off between the routing efficiency of proactive routing and the increasing traffic for maintaining the view of the zone.

3.5.1.4 Route Maintenance   Route maintenance is especially important in ad hoc networks, where links are broken and established as nodes move relatively to each other with limited radio coverage. In purely reactive routing protocols, until the new route is available, if any link fails, packets are dropped or, in some cases, delayed.

In ZRP, the knowledge of the local topology can be used for route maintenance. Link failures and suboptimal route segments within one zone can be bypassed. Incoming packets can be directed around the broken link through an active multihop path. Similarly, the topology can be used to shorten routes—for example, when two nodes have moved within each other's radio coverage. For source-routed packets, a relaying node can determine the closest route to the destination that is also a neighbor. Sometimes, a multihop segment can be replaced by a single hop. If next-hop forwarding is used, the nodes can make locally optimal decisions by selecting a shorter path.

3.5.2 Zone-Based Hierarchical Link State (ZHLS)

ZHLS is a hierarchical routing protocol, and it is a zone-based hierarchical LSR protocol that makes use of location information in a novel peer-to-peer hierarchical routing approach. The network is divided into zones that do not overlap. Aggregating nodes into zones conceals the detail of the network topology. Initially, each node knows its own position and therefore zone ID through GPS. After the network is established, each node knows the low-level (node level) topology about node connectivity within its zone and the high-level (zone level) topology about zone connectivity of the whole network. A packet is forwarded by specifying the hierarchical address—zone ID and node ID—of a destination node in the packet header.

When compared to other hierarchical protocols, there are no cluster heads in this protocol. The high-level topological information is distributed to all nodes (i.e., in a peer-to-peer manner). This peer-to-peer characteristic avoids traffic bottleneck, prevents single point of failure, and simplifies mobility management. Similar to ZRP, ZHLS is a hybrid reactive/proactive scheme. It is proactive if the destination is within the same zone of the source. Otherwise, it is reactive because it has to search the location to find the zone ID of the destination.

ZHLS requires GPS, which is not similar to ZRP and maintains a high-level hierarchy for interzone routing. Location search is performed by unicasting one location request to each zone. Routing is done by specifying the zone ID and the node ID of the destination, instead of specifying an ordered list of all the intermediate nodes between the source and the destination. Intermediate link breakage may not cause any subsequent location search. Since the network consists of nonoverlapping zones in ZHLS, frequency reuse is readily deployable in ZHLS.

3.5.2.1 Zone Map   The network is divided into zones under ZHLS. By making use of certain geolocation techniques such as GPS, each node knows its physical location; then, it can determine its zone ID by mapping its physical location to a zone map, which has to be worked out at the design stage.

The size of the zone depends on as the following characteristics: node mobility, network density, transmission power, and propagation characteristics. The partitioning can be based on simple geographic partitioning or on radio propagation partitioning. The geographic partitioning is much simpler and does not require any measurement of radio propagation characteristics, whereas the radio propagation partitioning is more accurate for frequency reuse.

3.5.2.2 Hierarchical Structure of ZHLS   Two levels of topology are defined in ZHLS: node level topology and zone level topology. A physical link exists if any two nodes are within the communication range. The node level topology (Figure 3.9) provides the information on how the nodes are connected together by these physical links.

For example, in Figure 3.9, if node "a" wants to send a data packet to node "i, " the data must pass through a–b–c–f. If there is at least one physical link connecting any two zones, a virtual link then exists.

To facilitate this hierarchical routing, two types of link-state packets (LSPs) are received by each node: node LSPs and zone LSPs. The node LSP of a particular node contains a list of its connected neighbors and is propagated locally within its zone. The zone LSP contains a list of its connected zones and is propagated globally throughout the network.

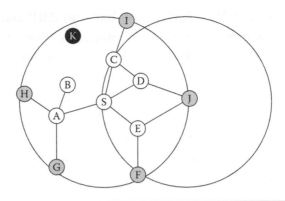

Figure 3.9   Node-level topology.

3.5.3 Distributed Dynamic Routing (DDR) Protocol

DDR is a hybrid approach based on zones that are not overlapped. In these types of protocols, each node is required to know only the next hop to all the nodes within its zone. This reduces routing information and bandwidth utilization. Each node keeps only the zone connectivity of its neighboring zones. The zone size again depends on the availability of neighbors. The zone names in DDR are assigned dynamically by some selected zone members. DDR avoids broadcasting by sending only the necessary information embedded in beacons to the neighboring nodes. The DDR also reduces maintenance cost and radio resource consumption overhead and leads to a stand-alone network. As there is no concept of network in DDR, there is no chance of a single point of failure.

The working of the DDR protocol can be divided into six phases: preferred neighbor election, intratree clustering, intertree clustering, forest construction, zone naming, and zone partitioning. Based on the information received at the beacon, each of these phases is executed. Each of these phases is executed based on information received in the beacon message.

A node can determine its neighbor based on ID number and the degree of neighboring nodes. After this, a forest is constructed by connecting each node to its preferred neighbor. Next, the intratree clustering algorithm is initiated to determine the structure of the zone and to build up the intratree routing table. This is then followed by the intertree algorithm to determine the connectivity with the neighboring zones. Then a zone name is assigned depending on ID number, node degree, or node stability during its lifetime in the zone.

By making use of different networks, a forest is constructed that contains a set of trees from a zone. The network is partitioned into a set of nonoverlapping dynamic zones constructed via gateway nodes. So the whole network can be seen as a set of connected zones. The size of the zone increases or decreases dynamically depending on some network failure such as node density/mobility, rate of connection/disconnection, and transmission process.

3.6 Summary

Mobile networks can be classified into infrastructure networks and mobile ad hoc networks according to their dependence on fixed infrastructures. In an infrastructure mobile network, mobile nodes have wired access points (or base stations) within their transmission range. The access points compose the backbone for an infrastructure network. In contrast, mobile ad hoc networks are autonomously self-organized networks without infrastructure support.

The routing architecture of self-organized networks can be either hierarchical or flat. In most self-organized networks, the hosts will be acting as independent routers, thus implying that routing architecture should conceptually be flat; that is, each address serves only as an identifier and does not convey any information about one host being topologically located with respect to any other node. Mobility management is not often necessary in a flat, self-organized network, as all of the nodes are visible to each other via routing protocols.

All existing routing protocols assume that all mobile hosts have the same properties based on the spirit of a self-organized network as a collection of "equal" peers opportunistically using each other's services to communicate. Although this is true in some circumstances, in some situations the network will include hosts with preponderant bandwidth, guaranteed power supply, and high-speed wireless links. Such hosts are referred to as superhosts.

QoS routing provides support to control effectively the total traffic that can flow into the network. QoS routing is a routing mechanism under which paths for flows are determined according to resource availability in the network as well as the QoS requirement of flows. QoS routing means that it selects routes with sufficient resources for the requested QoS parameters. Multicast routing sends a single copy of a data packet simultaneously to multiple receivers over a communication link that is shared by the paths to the receivers. The sharing of links in the collection of the paths to receivers implicitly defines a tree used to distribute multicast packets.

In contrast to unicast routing, multicast routing is a very useful and efficient way to support group communication. This is especially the case in self-organized networks where bandwidth is limited and energy is constrained. In addition, a self-organized network often consists of several cooperative work groups.

A proactive routing protocol is also called a "table-driven" routing protocol. In these protocols, nodes try to evaluate the routing tables continuously to maintain consistent, up-to-date routing information. Therefore, a source node can get a routing path immediately if it needs one.

Reactive routing protocols for mobile ad hoc networks are also called "on-demand" routing protocols. In a reactive routing protocol, routing paths are searched only when necessary. A route discovery operation invokes a route determination procedure. The discovery procedure terminates when either a route has been found or no route is available after examination for all route permutations. Hybrid routing protocols are proposed to combine the merits of both proactive and reactive routing protocols and overcome their shortcomings. Normally, hybrid routing protocols for mobile ad hoc networks exploit hierarchical network architectures. Proper proactive routing approaches and reactive routing approaches are exploited in different hierarchical levels.

Problems

3.1 Explain the classification of mobile ad hoc networks based on infrastructure.

3.2 Discuss briefly the design issues in developing a routing protocol.

3.3 List the factors that will make wireless links unidirectional in ad hoc networks.

3.4 Give suitable reasons why a traditional multicast routing protocol cannot be used in ad hoc networks.

3.5 Explain a proactive routing protocol with suitable examples.

3.6 Describe a wireless routing protocol with a neat diagram.

3.7 Discuss how routing is done using the destination-sequenced distance-vector protocol.

3.8 Explain how multipoint relays are used in the optimized link-state routing protocol.

3.9 How is message reduction done in fisheye state routing?

3.10 Explain a reactive routing protocol with an example.

3.11 Explain the ad hoc on-demand distance-vector routing protocol with a proper illustration.

3.12 Describe dynamic source routing used in ad hoc networks.

3.13 How is the temporally ordered routing algorithm used to route the packets in ad hoc networks?

3.14 Explain how the cluster-based routing protocol reduces overhead packets in ad hoc networks.

3.15 Discuss location-aided routing with an appropriate example.

3.16 With a suitable real-time example, explain the ant-colony-based routing algorithm.

3.17 Explain hybrid routing protocols with an example.

3.18 Explain the zone routing protocol with a neat diagram.

3.19 Discuss the zone-based hierarchical link-state routing protocol deployed in ad hoc networks.

3.20 Describe the scalable location updates routing protocol using a suitable example.

3.21 How does a distributed spanning trees-based routing protocol work well in ad hoc networks, and what are its disadvantages?

3.22 Explain distributed dynamic routing with a suitable example.

3.23 Explain the pros and cons of the DSDV routing protocol.

3.24 Discuss the pros and cons of the DSR protocol.

3.25 Describe the IMEP protocol in brief.

3.26 Discuss the benefits and disadvantages of the STAR protocol.

Bibliography

Abolhasan, M., T. Wysocki, and E. Dutkiewicz. 2004. A review of routing protocols for mobile ad hoc networks. *Elsevier Journal of Ad Hoc Networks* 2 (1): 1–22.

Adams, A., J. Nicholas, and W. Siadak. 2003. Protocol independent multicast—Dense mode (PIM-DM): Protocol specification (revised). IETF draft, February 2003.

Basagni, S., I. Chlamtac, V. R. Syrotivk, and B. A. Woodward. 1998. A distance effect algorithm for mobility (DREAM). *Proceedings of the Fourth Annual ACM/IEEE International Conference on Mobile Computing and Networking (MOBICOM'98),* Dallas, TX, 1998.

Bellur, B., R. G. Ogier, and F. L Templin. 2003. Topology broadcast based on reverse-path forwarding routing protocol (TBRPF). Internet draft, draft-ietf-manet-tbrpf-06.txt (work in progress).

Corson, M. S., and A. Ephremides. 1995. A distributed routing algorithm for mobile wireless networks. ACM/Baltzer *Wireless Networks* 1 (1): 61–81.

Corson, S., and J. Macker. RFC 2501: Mobile ad hoc networking (MANET): Routing protocol performance issues and evaluation considerations. Internet draft (draft-ietf-manet-issues-01.txt).

Das, S., C. Perkins, and E. Royer. 2000. Performance comparison of two on-demand routing protocols for ad hoc networks. *INFOCOM 2000,* pp. 3–12.

———. 2002. Ad hoc on demand distance vector (AODV) routing. Internet draft (draft-ietf-manetaodv-11.txt), work in progress.

Fenner, F., M. Handley, H. Holbrook, and I. Kouvelas. 2002. Protocol independent multicast—Sparse mode (PIM-SM): Protocol specification (revised). IETF draft, December 2002.

Ford, L. R., and D. R. Fulkerson. 1962. *Flows in networks.* Princeton, NJ: Princeton University Press.

Gerla, M. 2002. Fisheye state routing protocol (FSR) for ad hoc networks. Internet draft (draftietf-manet-aodv-03.txt), work in progress.

Geunes, M., U. Sorges, and I. Bouazizi. 2002. ARA: The ant-colony based routing algorithm for MANETs. *ICPP Workshop on Ad Hoc Networks* (IWAHN 2002), August, pp. 79–85.

Hass, Z. J., and R. Pearlman. 1999. Zone routing protocol for ad-hoc networks. Internet draft (draftietf-manet-zrp-02.txt), work in progress.

Hedrick, C. 1988. RFC 1058: Routing information protocol. Network Working Group, June 1988.

IETF MANET charter. http://www.ietf.org/html.charters/manet-charter.html

Jaquet, P., P. Muhlethaler, T. Clausen, A. Laouiti, A. Qayyum, and L. Viennot. 2001. Optimized link state routing protocol for ad hoc networks. *IEEE INMIC,* Pakistan.

Jiang, M., J. Ji, and Y. C. Tay. 1999. Cluster-based routing protocol. Internet draft (draft-ietfmanet-cbrp-spec-01.txt), work in progress.

Joa-Ng, M., and I-T. Lu. 1999. A peer-to-peer zone-based two-level link state routing for mobile ad hoc networks. *IEEE Journal on Selected Areas in Communications* 17 (8): 1415–1425.

Johnson, D., D. Maltz, and J. Jetcheva. 2002. The dynamic source routing protocol for mobile ad hoc networks. Internet draft (draft-ietf-manet-dsr-07.txt), work in progress.

Ko, Y. B., and N. H. Vaidya. 1998. Location aid routing (LAR) in mobile ad hoc networks. *Proceedings of the ACM/IEEE MOBICOM,* Oct. 1998.

Liu, C., and J. Kaiser. 2005. A survey of mobile ad hoc network routing protocols. University of Ulm technical report series, no. 2003-08, University of Ulm, Germany.

Lu, Y., W. Wang, Y. Zhong, and B. Bhargava. 2003. Study of distance vector routing protocols for mobile ad hoc networks. *Proceedings of the First IEEE International Conference on Pervasive Computing and Communications* (PerCom'03).

Meyer, R. A. *PARSEC user manual.* Los Angeles, CA: UCLA Parallel Computing Laboratory (http://pcl.cs.ucla.edu).

Moy, J. 1994. RFC 1584: Multicast open shortest path first (MOSPF). Network Working Group.

———. 1998. RFC 2328: OSPF version 2. Network Working Group.

Murthy, S., and J. J. Garcia-Luna-Aceves. 1995. A routing protocol for packet radio networks. *Proceedings of the First Annual ACM International Conference on Mobile Computing and Networking,* Berkeley, CA, pp. 86–95.

Nikaein, N., C. Bonnet, and N. Nikaein. 2001. HARP: Hybrid ad hoc routing protocol. *Proceedings of IST 2001: International Symposium on Telecommunications,* Tehran, Iran.

Nikaein, N., H. Laboid, and C. Bonnet. 2000. Distributed dynamic routing algorithm (DDR) for mobile ad hoc networks. *Proceedings of the MobiHOC 2000: First Annual Workshop on Mobile Ad Hoc Networking and Computing.*

NS-2, http://www.isi.edu/nsnam/ns/

OPNET Inc., http://www.opnet.com

Park, V., and S. Corson. 1997. Temporally ordered routing algorithm (TORA), version 1, functional specification. ETF Internet draft.

Pei, G., M. Gerla, X. Hong, and C. Chiang. 1999. A wireless hierarchical routing protocol with group mobility. *Proceedings of Wireless Communications and Networking,* New Orleans, LA.

Perkins, C. E., and T. J. Watson. 1994. Highly dynamic destination sequenced distance vector routing (DSDV) for mobile computers. *ACM SIGCOMM '94 Conference on Communications Architectures,* London.

Radhakrishnan, S., N. S. V. Rao, G. Racherla, C. N. Sekharan, and S. G. Batsell. 1999. DST: A routing protocol for ad hoc networks using distributed spanning trees. *IEEE Wireless Communications and Networking Conference,* New Orleans, LA.

Ramanathan, R., and M. Steenstrup. Hierarchically organized, multihop mobile wireless networks for quality of service. ACM/Baltzer *Mobile Networks and Applications* 3 (1): 101–119.

Royer, E., and C-K. Toh. 1999. A review of current routing protocols for ad hoc mobile wireless networks. *IEEE Personal Communications* 6:46–55.

Scalable Networks Inc. QualNet simulator software (http://www.scalable-networks.com).

Sinha, P., R. Sivakumar, and V. Bharghaven. 1999. CEDAR: A core-extraction distributed ad hoc routing algorithm. *IEEE INFOCOM.*

Tanenbaum, A. S. 1996. *Computer networks,* 3rd ed. Upper Saddle River, NJ: Prentice Hall.

Waitzman, D., C. Partridge, and S. Deering. 1988. RFC 1075: Distance vector multicast routing protocol. Network Working Group.

Woo, S-C., and S. Singh. 2001. Scalable routing protocol for ad hoc networks. *Wireless Networks* 7 (5): 513–529.

4

MULTICAST ROUTING
PROTOCOLS

4.1 Introduction

Wireless communication technology has been developed with two primary models. One is the fixed infrastructure-based model in which many of the nodes are mobile and connected through fixed backbone nodes using a wireless medium. Another model is the mobile ad hoc network (MANET). MANETs can be defined as a collection of mobile nodes (MNs) that are self-organizing and cooperative to ensure efficient and accurate packet routing between nodes (and to the base station also). There are no specific routers, servers, or access points for MANETs. Because of their speed and ease of deployment, robustness, and low cost, MANETs can be applied in fields such as military applications (i.e., to create a temporary network in the battlefield); search and rescue operations; temporary networks within meeting rooms; airports; vehicle-to-vehicle communication in smart transportation; establishing personal area networks connecting mobile devices like mobile phones, laptops, and smart watches; and other wearable computers, etc. To develop a routing protocol for wireless environments with mobility is very different from and more complex than developing those for wired networks with static nodes.

The main problem in mobile ad hoc networks is limited bandwidth and frequent change in the topology. Although there are lots of routing protocols that can be used for unicast and multicast communication within MANETs, it observes that any one protocol cannot fit in all the different scenarios, different topologies, and traffic patterns of mobile ad hoc networks applications. For example, proactive routing protocols are very useful for small-scale MANETs with high mobility, while reactive routing protocols are very useful for a large-scale MANETs with moderate or fewer topology changes.

Another set of routing protocols, known as hybrid routing protocols, will try to balance between the two—for example, proactive for neighborhoods and reactive for far away. Apart from this multicast is another category of routing protocol in MANETs that efficiently supports the group communication with the high throughput. The use of multicasting within MANETs has many benefits. It can decrease the cost of wireless communication and increase the efficiency and throughput of the wireless link between two nodes whenever multiple copies of the same messages are sent by the inherent broadcasting properties of wireless transmission. In place of sending the same data through multiple unicasts, multicasting decreases channel capacity consumption, sender nodes and router processing, energy utilization, and data delivery delay, which are important for MANETs.

If the mobile nodes in the MANET move too quickly, they have to be repaired in order to broadcast to achieve node-to-node communication. Each of the routing protocols has its advantages and disadvantages with regard to specific applications. The routing protocols should be designed in such a way that they minimize control traffic overhead; at the same time, they should be capable of rapidly linking failure and addition caused by node movements. Multicasting consists of concurrently sending the same message from one source to multiple destinations; it can be used in video conferencing, distance education, cooperative work, video on demand, replicated database updating and querying, etc.

4.2 Issues in Design of Multicast Routing Protocols

1. Scalability: A multicast protocol is scalable with constraints.
2. Address configuration: Different multicast groups have different addresses; reuse of a multicast group's address does not happen. Node movement causes synchronization of multicast addresses—a difficult task.
3. Multicast service support: Multicast participants should be able to join or leave the group on their own.

4. Security: The intruder should be stopped from joining an ongoing multicast session or receiving packets from other sessions.
5. Traffic control: The traffic should be efficiently distributed from central node to other members of the MANET.
6. QoS (quality of service): Multicast routing protocols should be designed in a way that satisfies a set of performance measures in terms of end-to-end delay, jitter, and available bandwidth.
7. Power control: Routing protocols must use less power as much as possible.
8. Multiple accesses: Most of the multicast protocols are designed for single source multicasting. In multiple source multicasting, each multicast source will induce its own overhead for multicast routing and waste network resources.

Depending on the topology used for communication, the multicast protocols can be classified as tree-based and mesh-based protocols. In tree-based protocols, only one route exists between a source and a destination and hence these protocols are efficient in terms of the number of link transmissions. There are two major categories of tree-based protocols: source tree based (the tree is rooted at the source) and shared tree based (the tree is rooted at a core node and all communication from the source nodes to the receiver nodes is routed through this core node). The shared tree-based multicast protocols are more scalable with respect to the number of sources; these protocols suffer under a single point of failure, the core node. On the other hand, source tree-based protocols are more efficient in terms of traffic distribution.

In mesh-based multicast routing, multiple routes exist between the source node and each of the receivers of the multicast group. A receiver node receives several copies of the data packets: one copy through each of the multiple paths. Mesh-based multicast routing protocols provide robustness in the presence of node mobility—however, at the expense of a larger number of link transmissions, leading to inefficient bandwidth usage. The mesh-based protocols are classified into source-initiated and receiver-initiated protocols depending on the entity (source or receiver) that initiates mesh formation.

4.3 Classification of Multicast Routing Protocols

4.3.1 Tree-Based Multicast Routing Protocols

4.3.1.1 Source Tree-Based Multicast Protocols In this section, we describe a representative protocol from each of the following major categories of source tree-based multicast routing protocols: (1) minimum hop based, (2) minimum link based, (3) stability based, and (4) zone based.

4.3.1.2 Minimum Hop-Based Multicast Protocols The minimum hop-based multicast routing protocols aim for a minimum hop path between the source node and every receiver node that is part of the multicast group. Each receiver node is connected to the source node on the shortest (i.e., the minimum hop) path, and the path is independent of the other paths connecting the source node to the rest of the multicast group. The multicast extension to the ad hoc on demand distance vector (MAODV) protocol is a classical example of minimum hop-based multicast routing protocols for MANETs. In this subsection, we describe the working of MAODV in detail.

> **Example: The Multicast Extension to the Ad Hoc On-Demand Distance Vector (MAODV) Protocol**
>
> *Tree formation (expansion) phase.* In MAODV, if a receiver node wants to join a multicast tree, it will join through a node that is a minimum hop path to the source. The receiver that wishes to join the multicast group broadcasts a route-request (RREQ) message. If the node that receives the RREQ message is not part of the tree, then the node broadcasts the message and establishes the reverse path by storing the state information, consisting of the group address, requesting the node ID and the sender node ID in a temporary cache.
>
> If the node that has received the message is a member of the multicast tree, the received node sends back a *route-reply* (*RREP*) message on the shortest path to the receiver. In the RREP message, the information about the number of hops from the node itself to the source is revealed. The receiver that wants to join the group receives several *RREP* messages and selects the member node that lies on the shortest path to the source. The receiver node sends a *multicast activation* (*MACT*) message to the selected

member node along the chosen route. The route setup happens when the member node and all the intermediate nodes in the chosen path update their multicast table with state information from the temporary cache.

Working. In Figure 4.1, we illustrate tree formation (expansion) under the MAODV protocol using an example. Here, the multicast tree is already established between the source node, S, and the two receivers, R1 and R2, of the multicast group. A node is considered a multicast tree node if it is a source, receiver, or intermediate node of the multicast tree.

Now, consider a new member, node R3, joining as the receiver of the multicast group. To become part of the multicast tree, R3 broadcasts a route request (*RREQ*) control packet to its neighbors. If a neighbor node is a multicast tree node, it does not further propagate the *RREQ* packet; otherwise, it broadcasts the packet to its neighbors. The multicast tree node that receives an *RREQ* packet waits for a certain time period to receive any more *RREQ* packets and then responds back with a route reply (*RREP*) packet on the shortest (minimum hop) path to the initiator (R3). In the *RREP* packet, the tree node also includes the number of hops between itself and the source.

In our example, intermediate tree node I1 (on the shortest path from S to R2) and the receivers R1 and R2 respond back with *RREP* packets to R3. The number of hops on the shortest path from R3 to each of R1 and R2 is 4, whereas the number of hops from I1 to R3 is 3. Also, the number of hops from S to R1 and R2 on the shortest path is, respectively, 2 and 3; the number of

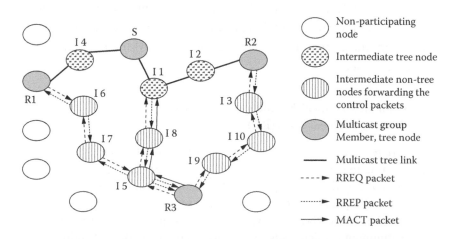

Figure.4.1   MAODV: receiver joining a multicast tree.

hops from S to I1 is 1. Considering all these, the shortest path from R3 to the source S would be the path that goes through the intermediate tree node I1. Hence, R3 decides to join the multicast tree through I1 and sends a multicast activation (*MACT*) message to I1. The intermediate nodes I5 and I8 that forwarded all of the three control packets (*RREQ, RREP,* and *MACT*) now become part of the multicast tree.

Tree maintenance phase. The tree is maintained by the expanding ring search (ERS) approach, using the *RREQ, RREP,* and *MACT* messages. The downstream node of a broken link is responsible for initiating ERS to issue a fresh *RREQ* for the group. This *RREQ* contains the hop count of the requesting node from the multicast source node and the last known sequence number for that group. It can be replied to only by the member nodes whose recorded sequence number is greater than that indicated in the *RREQ* and whose hop distance to the source is smaller than the value indicated in the *RREQ*.

4.3.1.3 Minimum Link-Based Multicast Protocols   The minimum link-based multicast protocols try to have a minimum number of links overall in the multicast tree connecting a source node to all the receiver nodes of the multicast group. Such a tree would use the bandwidth efficiently and facilitate simultaneous use of the wireless channel for several node pairs whose communications will not interfere with each other. The bandwidth efficient multicast routing protocol (BEMRP) is an example of a minimum link-based source-tree multicast routing protocol, since the protocol tries to minimize the number of additional links that get incorporated when a new receiver node joins an already existing multicast tree. In this subsection, the working of BEMRP is discussed in detail.

Example: Bandwidth-Efficient Multicast Routing Protocol (BEMRP)

Tree formation (expansion) phase. According to BEMRP, a node wanting to join the multicast group opts for the nearest forwarding node in the existing tree, rather than choosing a minimum hop count path from the source of the multicast group. Because of this, the number of newly added links in the multicast tree is minimum, which leads to savings in the network bandwidth. Multicast tree construction is receiver initiated. If a node wishes to join the multicast group as a receiver, the node starts flooding the *join* control packets targeted toward the nodes that are currently members of the multicast tree. After receiving the first *join* control packet, the member node waits for some amount of time before sending a *reply* packet. The member node sends a *reply* packet on the path traversed by the *join* control packet, with the minimum number of intermediate forwarding nodes. The node that wants to join as receiving node collects the *reply* packets from different member nodes and sends a *reserve* packet on the path that has a minimum number of forwarding nodes from the member node to itself.

Working. The working of BEMRP is illustrated using an example shown in Figure 4.2. Let us consider a minimum link multicast tree connecting the source node S and the two receivers R1 and R2 of the multicast group (the darkened links shown in Figure 4.3 form the multicast tree). Similarly to MAODV, a tree node can be either a multicast group member (source and receiver nodes) or an intermediate node in the tree. If a third node, R3, wants to join the multicast group, it broadcasts the *join* control packets in its neighborhood and the packets get forwarded until they are received by a tree node. After receiving the first *join* control packet, the tree node waits for a while and sends back a *reply* packet on the path that has the minimum hop count to the initiator of the *join* control packet (R3).

However, unlike MAODV, the number of hops from the responding tree node to the source of the multicast group is not

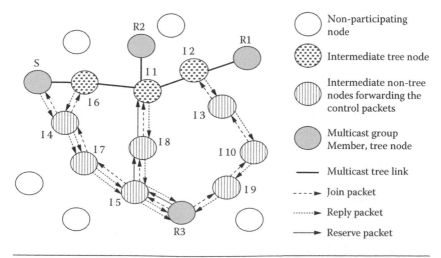

Figure 4.2 BEMRP: receiver joining a multicast tree.

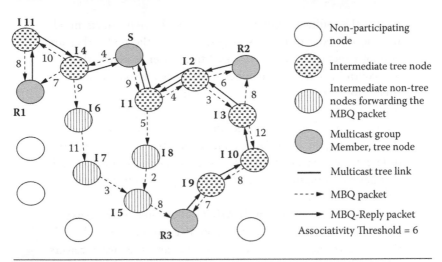

Figure 4.3   ABAM: source-initiated tree construction.

included in the *reply* packet and only the hop count of the path from the tree node to the receiver is included. The receiving node R3 collects all the *reply* packets and decides to join the multicast tree through the closest tree node (I1) so that the number of new links (three, in this case) that will be added to the multicast tree will be the minimum. The node R3 sends a *reserve* packet to the chosen intermediate tree node and the nodes of the links traversed by this *reserve* packet during path detection are now part of the minimum link-based multicast tree.

Note that if node R3 had chosen to send the *reserve* packet to the source S directly by responding to its *reply* packet, R3 would have been connected to the source on a minimum hop path. However, such a path to the already existing multicast tree would create an addition of four new links to the tree and hence R3 prefers to go through I1 (which has resulted in only three new links added to the tree). The trade-off is that the number of hops from the source S to the receiver R3 is now five and this is larger than the minimum number of hops between S and R3 in the network.

Tree maintenance phase. BEMRP is a hard-state-based tree maintenance approach, since it has to provide more bandwidth efficiency (i.e., a member node transmits control packets only after a link breaks). BEMRP uses two schemes to recover from link failures: the *broadcast-multicast scheme*, in which the upstream node of the broken link is responsible for finding a new route to the previous downstream node; and the *local rejoin scheme*, in which, if any broken link is there, it tries to rejoin the multicast group using a limited flooding of the *join* control packets.

4.3.1.4 Stability-Based Multicast Protocols Stability-based multicast protocols aim for a long-living tree connecting the source node to the receiver nodes of the multicast group. Each receiver node, at the time of joining the tree, selects the most stable path to the source node. This would minimize the number of tree reconfigurations. To find the stable path, it uses metrics that are a measure of the longevity of the links in the network. Metrics such as predicted link expiration time (LET), link affinity, and associativity ticks have been used by the routing protocols for determining stable paths as well as stable trees. Here, one such multicast routing protocol called the associativity-based ad hoc multicast (ABAM) routing protocol is discussed.

Example: Associativity-Based Ad Hoc
Multicast (ABAM) Routing Protocol

Formation (expansion) phase. This is the stability of link associativity ticks, which is the number of beacons periodically received from that neighbor since the link was formed. Each node stores the value of the associativity ticks with its neighbors. Multicast tree construction is source initiated and it can be repaired or expanded through receiver-initiated broadcast queries. The tree construction phase is initiated by the source node by broadcasting a multicast broadcast query (*MBQ*) message in the network to inform all prospective receiver nodes. The intermediate node appends associativity ticks if it receives the *MBQ* message and then rebroadcasts it.

Upon receiving several *MBQ* messages through different paths, a receiver node of the multicast group selects the most stable path and sends an *MBQ-reply* packet along the selected path. The most stable path is the path with the largest proportion (i.e., percentage) of stable links. If the value of the associativity ticks is greater than or equal to the associativity threshold, calculated based on the node velocity and transmission range per node, then the link is called stable. In case of a tie (i.e., if two or more paths have the same largest proportion of stable links), then the minimum hop path among the contending paths is chosen. After receiving *MBQ-reply* packets from each receiver of the group, the source sends *multicast MC-setup* messages to all receivers in order to establish the multicast tree.

Working. The working examples presented so far in the previous sections (on MAODV and BEMRP) have been receiver initiated. Receiver-initiated tree repair and expansion of ABAM would also be similar. In the case of ABAM, the route selection metric adopted by the tree node receiving the *join query* packets would be based on the associativity ticks of the links traversed by the *join query* packets. The tree node chooses the path with the largest proportion of stable links and sends a *join reply* packet. A working example that is source initiated and based on the concept of choosing paths with the largest proportion of stable links is presented here. As shown in Figure 4.3, the source node S initiates a broadcast query reply cycle of the *MBQ* packets. In the diagram, associativity tick value is represented as link weight. The associativity threshold value is assumed to be six. If the associativity tick value for a link is greater than or equal to the threshold, the link is said to be "stable"; otherwise, it is unstable. If any multicast receiver node receives *MBQ* packets across several paths, it calculates the proportion of stable links on each of these paths and chooses the path with the largest proportion of stable links.

Tree maintenance phase. Tree maintenance is by using a *local query reply* cycle. If any link breaks, it attempts to fix a route to the receiver node by broadcasting a *local query* message with TTL value of one. When the receiver node receives the *local query* message, it responds with a *local reply* message. The upstream node then sends the *MC-setup* message to the receiver. The responsibility of fixing the route to the receiver is transferred if the upstream node cannot find a route to the receiver; then it transfers the responsibility of fixing the route to its immediate upstream node on the path from the receiver to the source. With TTL value set to two, this upstream node then initiates a broadcast of the *local query* message. This procedure is continued until the timer at the receiver node expires and it broadcasts a *join query* message to join the multicast group.

The *join query* message is broadcast by a newly joining receiver node or a receiver node that got cut off from the multicast tree and could not join the tree using a *local query reply* cycle. The *join query* message propagates until it reaches a tree node. During its propagation, the forwarding nodes append their node ID and the associativity tick values with the downstream node from which the message was received. When a tree node receives a specific *join query* message (identified using a sequence number) for the first time, it waits for a while to receive the *join query* messages along different paths, chooses the path with the largest proportion of stable links, and sends a *join reply* message on the selected stable path. The receiver node confirms its participation in the multicast session by sending a *reserve* message on the path traversed by the *join reply* message.

4.3.1.5 Multicast Zone-Based Routing Protocol (MZRP)   MZRP is the multicast extension of the unicast zone routing protocol (ZRP), a hybrid of both proactive and reactive routing strategies. A zone in the network comprises nodes that are within two or three hops from each other. Multiple zones exist in the network and often these zones overlap with each other. A border node is the node that is part of more than one zone. ZRP makes use of proactive routing for intra-zone communication and, for interzone communication, it uses the combination of proactive and reactive routing protocols. For example, if the source and destination nodes are in the same zone, then pro-active routing protocol is used. Otherwise, if they are in different zones, the source node has to utilize the proactive routing protocols implemented in their respective zones and a reactive routing protocol implemented for interzone communication. ZRP does not depend on any specific proactive and reactive routing protocol for intrazone and interzone communication. Similarly, MZRP does not depend on any underlying unicast routing protocol. The proactive routing mechanism within a zone is implemented through periodic beacon exchange and the reactive routing mechanism to communicate across different zones is realized through on-demand flooding and multi-cast tree construction.

Node advertisement and zone initialization. Every node advertises to the other zone members by broadcasting an *advertisement* packet, and its propagation is controlled using a time-to-live (*TTL*) value that is usually set to the zone radius. The nodes that receive the *advertisement* packet make an entry for the source in their *zone routing table* and update their *neighbor table* with the upstream node that sent the *adver-tisement* packet. ZRP operates in a soft state (i.e., nodes remove the route entry of a particular node that fails to send periodic *advertisement*

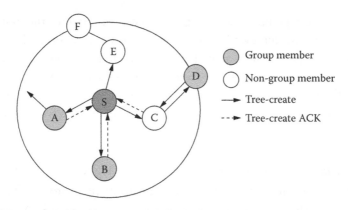

Figure 4.4   Multicast tree creation inside a zone.

packets to indicate its presence in the zone). Thus, topology changes within a zone are often localized and not broadcast to other zones.

Multicast tree creation inside a zone. Multicast tree creation within a zone (illustrated in Figure 4.4) is done as follows: The initial node or the source node broadcasts a *tree-create* packet, which contains a session ID, with a *TTL* value set to the radius of the zone. After receiving the *tree-create* packet, any node creates a multicast route entry with an empty list of downstream nodes and the upstream node is set to the node that sent the *tree-create* packet. Any node within the zone that wants to be a receiver of the multicast session responds with a *tree-create-ACK (acknowledge)* packet and the packet travels back to the source on the reverse path traversed by the *tree-create* packet. On this path any intermediate node that receives the *tree-create-ACK* packet for a multicast session for which an entry has already been created in its routing table updates the downstream node list by adding the ID of the neighbor node from which the *tree-create-ACK* packet was received.

4.3.1.5.1 Extension of the Multicast Tree to the Entire Network   If the multicast tree is to be extended to the entire network (refer to Figure 4.5), the source node at a *tree-propagate* packet will be sent by the source node to the border nodes of its own zone. A border node that receives a *tree-propagate* packet creates a route entry for the session in its multicast table and initiates the zone-wide broadcast of a *tree-create* packet. Any node interested in joining the zone responds with a *tree-create-ACK* packet that is duly forwarded by the border

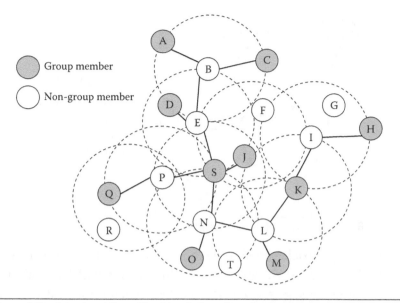

Figure 4.5 Extension of the multicast tree to the entire network.

node to the multicast source. A network-wide TTL value is specified indicating the number of zones it can be forwarded to. A border
node decrements this *TTL* value by one and, when it reaches zero,
the *tree-propagate packet* is sent to the other border nodes within that
zone. After the construction of a multicast tree, the source node starts
sending data on the tree, which will be propagated through the nodes
across zones, according to the downstream list maintained by intermediate nodes.

4.3.1.5.2 Zone and Multicast Tree Maintenance Multicast tree maintenance is by soft state. Periodically, for every interval, the source
node sends a tree-refresh packet. Any node that gets disconnected
from the tree sends a join packet, identifying the multicast session, to
all of its zone nodes. Any other node of the multicast tree that has not
disconnected responds with a *join-ACK* packet and adds the neighbor
node that sent the *join* packet to the list of downstream nodes. A
similar procedure is adopted at all the nodes that receive the *join* and/
or *join-ACK* packets.

Figure 4.6 (adapted from reference 1) illustrates the rejoin process
within a zone. After sending the *join* packet, if any disconnected node
does not receive the *join-ACK* packet within a limited time, it sends

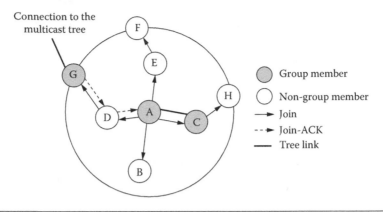

Figure 4.6   Zone and multicast tree maintenance.

a *join-propagate* packet to all of its border nodes, which in turn send *join* packets to all their zone nodes. This process will be continued until a border node gets a *join-ACK* packet, which is forwarded to the disconnected member node.

A receiver node disconnects from the multicast session by sending a *tree-prune* message to its upstream node in the tree. The upstream node will remove the receiver node from its list of downstream nodes for the tree; if the list becomes empty after the removal and the node is not an interested receiver node (i.e., has been just an intermediate tree node), then the node sends a *tree-prune* message to its upstream node further up in the tree

4.3.1.6 Shared Tree-Based Multicast Protocols   In shared tree-based protocols, a single tree is constructed where the central control point is called the rendezvous point (*RP*). The mobility in MANETs will be typically in two levels, with a set of nodes moving fast and the rest of the nodes moving relatively very slowly. The *RP* is often chosen to be the slowest moving node in a MANET so that the shared tree can be maintained irrespective of the mobility of the source nodes. The *RP* is connected to the receiver nodes of the multicast session and the source nodes send the data packets to the *RP* based on the underlying unicast routing protocol.

Still, shared tree-based protocols are known to yield a lower throughput compared to per-source tree-based protocols. A compromise solution called "adaptive-tree multicast" is applicable wherein if the receivers request the source, the source node may construct its own rooted tree to deliver the packets on the shortest path. In this section, the study of the shared-tree wireless multicast (ST-WIM) protocol, ad hoc multicast routing protocol utilizing increasing ID numbers (AMRIS), and the ad hoc multicast routing protocol (AMRoute) as representatives of the cluster-based, session-specific, and IP (Internet protocol) multicast session-based protocols.

4.3.1.7 Session-Specific Ad Hoc Multicast Routing Protocol Utilizing Increasing ID Numbers (AMRIS) AMRIS provides a unique session-specific multicast session member ID (*msm id*) to each participant. The *msm id* indicates the logical height of a node in the multicast delivery tree rooted at the sender that has the smallest *msm id* (*Sid*) in the tree. All other nodes in the tree have an *msm id* that is higher than that of its parent. In case of a multiple sender environment, the *Sid* is elected among the senders. AMRIS uses the underlying *MAC* layer beaconing mechanism to detect the presence of neighbors.

Tree initialization phase. During the tree initialization, the *Sid* broadcasts a *new-session* message (containing the *Sid, msm id,* and other routing metrics) in its neighborhood. If a node receives the *new-session* message, it updates the *msm id* in the message with a newly computed larger value that is also used to identify the node in the tree. If more than one *new-session* message is received from several neighbors within a random jitter amount of time, then the message with the best routing metric is selected and updated with a newly computed *msm id* value. The updated *new-session* message is then rebroadcast. This strategy can be used to prevent broadcast storms. Note that the newly computed *msm ids* are not consecutive and the gaps can be used to repair the delivery tree locally.

To join the multicast group, a downstream node X sends a unicast *join-request* message to a randomly chosen neighbor node that is also a potential parent node (having a lower *msm id*)—say, Y. If node Y already has become part of the multicast tree, then Y sends a *join-ACK* message to X. Otherwise, Y forwards the *join-request* message to a set of potential parents of itself. This process continues until the *join-request* message reaches a parent node that is part of the multicast delivery tree. A *join-ACK* message is sent by the parent node to node X, which confirms the participation of node X in the multicast session by sending back a *join-conf* message to the parent node.

Tree maintenance phase. Upon link failure, the downstream node (with a relatively higher *msm id*) is responsible for rejoining the multicast tree. Before broadcasting a *join-request* message, the downstream node of a broken link attempts to find potential parent nodes by locally repairing the route using an expanded-ring search process. The *join-request* message is broadcast with a *TTL* value that restricts the number of hops the message can propagate. If a node that wants to rejoin a muticast tree has no valid *msm id*, it calculates *msm id* for itself based on the *msm id*s of its neighboring node's *msm id* and then joins the multicast tree through the branch reconstruction process explained before. The beacon update period at the *MAC* layer has to be chosen properly to avoid detection of microterm breakages that can unnecessarily trigger the branch reconstruction process and incur a lot of control overhead.

4.3.2 Mesh-Based Multicast Routing Protocols

4.3.2.1 Source-Initiated Mesh-Based Multicast Protocols   A mesh is a set of nodes in the network such that all the nodes in the mesh forward multicast packets via scoped flooding. As stated before, mesh-based protocols are more robust to link failures than tree-based protocols. Mesh-based protocols can be either source initiated or receiver initiated. In most of the cases, the forwarding mesh in source-initiated protocols is a union of per-source meshes, while receiver-initiated mesh protocols form a single shared mesh for all the sources. Here, we discuss the source-initiated mesh-based multicast routing protocols and, in the next section, we discuss the receiver-initiated mesh-based protocols. In the category of source-initiated mesh-based multicast routing protocols, we discuss the well known on-demand multicast routing protocol (ODMRP) along with its extensions to handle high mobility and low node density (i.e., sparse networks).

4.3.2.1.1 On-Demand Multicast Routing Protocol (ODMRP)   ODMRP is a mesh-based multicast routing protocol based on the notion of a forwarding group (shown in Figure 4.7)—a set of nodes that forward data on the shortest paths between any two multicast members. Multicast group membership and routes are established and

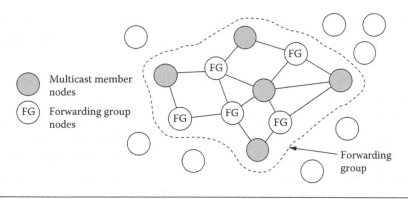

Figure 4.7 Concept of forwarding zone.

updated by the source on an on-demand basis. This leads to reduction in channel/storage overhead and an increase in scalability. A soft-state approach is used for mesh maintenance and member nodes are not required to send leave messages explicitly while quitting a group. A performance comparison of the major ad hoc multicast routing protocols shows ODMRP to be the most advantageous and preferred protocol in mobile wireless networks. ODMRP can also operate independently as an efficient unicast routing protocol.

ODMRP operates through a request phase and a reply phase. The source nodes, which are not aware of routes or membership, broadcast a *join-data* packet. When any node receives the *join-data* packet for the first time, its routing table is updated by storing the upstream node ID and rebroadcasts the packet. The multicast receiver, after receiving a nonduplicate *join-data* packet, creates and broadcasts a *join-reply* packet in its neighborhood.

When a node receives the *join-reply* packet, it checks if it is listed as the next node ID in the packet. If it is, the node is located on the path to the source and becomes part of the forwarding group. An FG (forwarding group) flag is set in its routing table and the node broadcasts its own *join-reply* packet. In this way, the *join-reply* packet will be forwarded by the FG member nodes until it reaches the multicast source on the shortest path. A forwarding group is built as a result of this source-receiver route construction and update process. If any node wants to leave a multicast group, it can respectively stop sending the *join-query* and *join-reply* packets meant for that group. If an FG member node does not receive a *join-reply* packet before a time-out period, the node demotes itself to a nonforwarding node of the mesh.

4.3.2.2 Receiver-Initiated Mesh-Based Multicast Protocols Receiver-initiated mesh protocols are more robust to node mobility as they attempt to maintain a shared mesh involving all the source nodes of the multicast group. The multicast source nodes forward packets on the reverse shortest path from the receiver nodes to the source. This section reviews the work of the receiver-initiated multicast mesh protocol: the core-assisted mesh protocol (CAMP).

4.3.2.2.1 Core-Assisted Mesh Protocol (CAMP) In CAMP the nodes can be categorized into three types: duplex, simplex, and nonmembers. Duplex nodes will be similar to regular mesh nodes with functionality the same as mesh routing protocols. Simplex members are able to forward packets from the source nodes to the rest of the mesh, but they cannot respond for any membership query packets. CAMP maintains one or more core nodes per mesh. A node interested in joining the multicast mesh first queries its neighbor nodes to see if any of them are part of the mesh. If none of the neighbors are part of the mesh, the node will send the *join-request* messages by flooding targeted toward the core nodes of the mesh. Upon receiving a nonduplicate *join-request* message, a duplex member node, which is also called core node, responds with a *join-ACK* message that is propagated back to the initiator of the *join-request* message. If the core nodes are not able to provide efficient dissemination paths, CAMP lets core nodes leave the multicast mesh. A core node leaves the multicast group by broadcasting a *quit* notification message to its neighbors.

A receiver node periodically checks whether the multicast data packets traverse the reverse shortest path back to the source. If this is not the case, the receiver node sends a *heart-beat* or the *push-join* message along the reverse shortest path to the source. A member node (including the receiver node) forwards the *heart-beat* message to its successor node on the reverse shortest path if the latter is already a member of the mesh; otherwise, the member node sends a *push-join* message to the neighbor successor node asking it to join the multicast mesh and waits for an *ACK* from that node. Duplex members respond with a regular *ACK*, while simplex members send an *ACK-simplex* message. If no *ACK* is received within a certain time, the *push-join* message is propagated further until a member node that is directly connected to the source is reached.

The member nodes of a mesh are allowed to select their "anchor" nodes, which are required to rebroadcast any nonredundant data packets received. The member nodes can periodically refresh their "anchor" nodes by broadcasting updates. Any neighbor node that is not interested can deny the "anchor" request or discontinue from that role.

CAMP requires the support of the underlying unicast routing protocol to provide route updates. Because of this, CAMP prefers to coexist with an underlying unicast routing protocol called wireless routing protocol (WRP), which marks a subset of destinations as unreachable during periods of network reconvergence. CAMP piggybacks its control messages onto the back of WRP updates and the control overhead is bound to increase exponentially with traffic. We cannot use CAMP directly with unicast routing protocols based on the Bellman–Ford algorithm; extensions are needed to work with on-demand routing protocols.

4.3.3 Source-Based Multicast Routing Protocol

The source-based multicast routing protocol operates in a loop-free manner and tries to minimize both routing and storage overhead in order to provide robustness to host mobility, adaptability to wireless channel fluctuation, and optimization of network resource use. SRMP is a mesh-based, multicast routing protocol; a mesh structure is established on demand to connect group members, and a multicast mesh provides at least one path from each source to each receiver in the multicast group. Route selection is established through a multicast mesh, started at the multicast receiver. The concept of FG (forwarding group) nodes is used during mesh establishment. The FG is a set of selected nodes responsible for forwarding multicast data between any number of pairs.

4.3.3.1 FG Node Selection The choice of FG nodes and how they are selected and maintained is the key challenge in efficient multicasting. SRMP achieves a compromise between the size of the selected nodes and the availability and stability of the selected paths. SRMP applies efficient FG node selection criteria through defining four metrics: association stability, link signal strength, battery life, and link availability.

- Association stability: To measure how long the node is stable with respect to its neighbor, this metric is used. It was first introduced in the ABR protocol and is known as the degree of association stability. Association stability is calculated by each node with respect to each neighbor through the use of an *associativity ticks* field stored in the node's *Neighbor_Stability_Table*. It is incremented each time the node receives a beacon indicating a neighbor's existence. A node is considered stable with respect to a neighbor when the accumulated *associativity ticks* value corresponding to this neighbor fulfills a predefined threshold.
- Link signal strength: To measure the signal strength between each node and its neighbors this metric is used, which indicates connectivity strength. SRMP uses this metric to select links that offer stronger connectivity between nodes. Signal strength is calculated according to the level of strength the beacon is received and classified as weak or strong.
- Battery life: This metric calculates the current battery power, which is a decreasing function of time and processed packets.
- Link availability: This metric is used during path selection; here, prediction-based link availability estimation is used.

The following data structures are used in SRMP:

- The *Neighbor_Stability_Table* gathers continuous node-neighbor information.
- The *Multicast_Message_Duplication_Table* that identifies each received Join-request or data packet,
- The *Multicast_Routing_Cache* stores all possible routes from each node to each multicast group.
- The *Receiver_Multicast_Routing_Table* is maintained at each receiver for each multicast group, and stores the used route between each receiver and each source

Neighbor	Type	Associativity Ticks	Signal Strength	Link Availability

Table 3: Neighbor_Sability_Table

Source ID	Sequence Number	Type

Table 4: Multicast_Message_Duplication_Table

Group ID	Source ID	Route to Source	Timer

Table 5: Receiver_Multicast_Routing_Table

Group ID	Type	Route to Receiver	Timer

Table 6: Multicast_Routing_Cache

4.3.3.2 Operation Similarly to all on-demand routing protocols, the request phase and a reply phase are involved in the operation of a protocol. The request phase involves a route discovery phase to find a route to reach the multicast group. During the reply phase, different routes to multicast groups are set up through FG node selection and mesh construction. In the request phase, a source node that is not a member of the multicast group wants to join the group; it starts a route discovery procedure by broadcasting a *join-request* packet to a neighbor. This *join-request* packet contains the ID of the source node in a source-ID field, the multicast group-ID and destination ID fields, and sequence number field.

Reply phase and mesh construction. When a *join-request* packet is received by the corresponding receiver, it checks for stability among its neighbors, including associativity ticks' signal strength and link availability. After satisfying all the metrics for the predefined thresholds, the receiver selects a neighbor as FG node and sets it as a member in the neighbor-stability table. The receiver sends a *join-reply* packet to this FG node, storing the multicast group ID in a source field and the ID of the requesting field as destination ID field. A source route also accumulates during *join-reply* propagation in a *route record* field in the packet.

An FG node receiving a *join-reply* first creates an entry to the multicast group in its *Multicast_Routing_Cache,* setting its state as FG node and copying the reversed accumulated route in the received *join-reply*. It then performs the same previous steps for selecting FG nodes among neighbors. This process continues until reaching the source, constructing a mesh of FG nodes that connect group members. A source receiving a *join-reply* packet creates an entry to the multicast group in its *Multicast_Routing_Cache*.

More than one *join-reply* may be received by the source for the same multicast group. Hence, multiple routes can be stored for the same multicast group. After mesh creation, new *join-request* packets may receive replies from any FG member node having unexpired routes to the requested multicast group. In this case, the FG node sends the *join-reply* following the same previous selection among neighbors.

Data transmission. The shortest path route is selected to transmit data. If more than one shortest path routes are found, the freshest route is selected. Data packets carry in their headers the selected route indicating the sequence of hops to be followed. Each FG node receiving a data packet forwards this packet if it stores in its cache at least one valid route toward the multicast group and the packet is not duplicated. This leads to an attractive feature in SRMP: preventing packets transmission through stale routes and minimizing traffic overhead. The process continues until reaching all multicast receivers.

A multicast receiver receiving a data packet for the first time creates an entry in its *Receiver_Multicast_Routing_Table*. To guarantee data transmission to all multicast receivers, nodes duplicate transmission if the selected route leads directly to the multicast group. We define duplication in transmission as selecting one more route following same previous criteria and transmitting data to both routes.

Maintenance. Route maintenance deals with reporting and recovering routing problems, keeping the lifetime of a route as long as possible. For this purpose, SRMP addresses four mechanisms: providing multicast mesh refreshment, link breaks detection and repair, continuous node-neighbor information, and pruning allowing any node to leave the group. Two new messages are introduced: the *multicast-RERR (route error)* message and the *leave group* message.

Neighbor existence mechanism. SRMP uses MAC layer beacons to provide each node with neighbors' existence information. Upon reception of neighbors' beacons, creating or updating *Neighbor_Stability_Table* entries takes place via incrementing the associativity *ticks* and setting the *signal strength* according to the level of strength of the beacon that is received. In addition, *link availability* is updated by continuous prediction for links' availability toward neighbors. If no beacons are received by a node from a neighbor up to a certain period of time, the node indicates the neighbor's movement and updates its stability table fields toward it.

Mesh refreshment mechanism. This follows a simple mechanism making use of data packet propagation and requiring no extra control overhead. Each time the source transmits a data packet, in its cache it updates the timer of the used route. Typically, an FG node forwarding this packet scans the packet header and refreshes in its cache the corresponding route entry timer. Furthermore, a multicast receiver scans the header of each received data packet, refreshing its corresponding table entry timer to the source. Periodically, each node checks its timers and purges out expired multicast group entries, preventing stale route storage. In addition, it checks its neighbor table, deleting from its cache routes to multicast groups for which it possesses no more members.

Link repair mechanism. With the help of MAC layer support, SRMP detects link failures during data transmission. In this case, two mechanisms are addressed: maintaining routes when a link fails between two FG nodes and maintaining routes when a link fails between a multicast receiver and an FG node. In fact, mesh reconfigurations are not needed if the stability characteristics together with high battery life paths are valid throughout the lifetime of the multicast communications. A link's failure occurs between two FG nodes; the node detecting failure reports it to the original source following the same procedure of link failure recovery in the DSR protocol. First, it generates a *multicast-RERR* packet indicating the broken link in a *broken link* field in this packet. Then, it deletes from its cache any routes containing the broken link. In turn, nodes on the way to the source that receive this packet clean their caches from all routes containing the broken link.

When links fail between an FG node and a multicast receiver, an alternative approach is applied. Simply, the FG node detecting failure deletes the receiver membership from its *Neighbor_Stability_Table*. When the FG node possesses no more members for the multicast group, it deletes routes to this group from its cache and sends to all its neighbors a *multicast-RERR* packet reporting the failure. The *broken link* field in this packet stores the link between the failure detector FG node and the multicast group. Each neighbor, receiving this packet, cleans its cache from routes containing the broken link. The process is repeated until all member nodes in the mesh are visited.

Pruning mechanism. An effective pruning mechanism is implemented by SRMP, allowing a member node to leave the multicast session. It mainly deals with two cases: FG node pruning and multicast receiver pruning. A multicast source wishing to leave a multicast group simply stops transmitting data to this group, deleting from its cache entries concerning this group.

If a node wishes to leave a multicast group, it sends a *leave group* message to its member neighbors, deleting from its table all entries corresponding to this group. The *leave group* message carries the ID of the multicast session in a *multicast group ID* field and the ID of the member neighbor to which the message is sent in a *neighbor ID* field. The neighbor node receiving this message cancels in its turn the receiver membership from its *Neighbor_Stability_Table*. When this node has no more members for a multicast group, it sends in its turn a *multicast-RERR* message to its member neighbors following previous procedure in link failure.

SRMP performs better compared to other protocols by introducing no extra control overhead and efficient repair and pruning mechanisms.

4.4 QoS Routing

In a wireless network, as communication happens through radio packets, direct communication is allowed only between direct nodes, if the nodes are to be communicated which are distant, communication should happen through multihop nodes. As the real-world applications need QOS support, QOS routing is helpful in multicasting. Since network topology changes very frequently in MANETs, QOS routing in MANETs is difficult.

Another challenge in QoS for real-time applications is associated with design of the MAC protocol. Because the topology changes dynamically, it is difficult to provide reservation, central controller, etc. The requirement for QoS in MANETs is to find a route through the network that is capable of supporting a requested level of QoS. When existing network topology changes, new routes can support existing QoS and respond to the changes in available resources.

QoS in MANETs is highly dependent on routing and medium access control. A CDMA/TDMA channel model is used in a MAC layer for most of the implementations of unicast routing with QoS. It is difficult to implement CDMA/TDMA in a real network due to issues of code and synchronization between the nodes. Even though highly synchronized solutions are used in MANETS, they may fail when nodes become mobile and they will also be expensive.

4.4.1 Multicast Routing in QoS

A node that has data to send starts a session by broadcasting a session initiation as a quality of service route request (QRREQ) with TTL greater than zero. The intermediate node rebroadcasts the QRREQ, if it has bandwidth, until TTL is equal to zero. The destination node receives QRREQ and sends a QoS route reply (QRREP) to the source.

Forward group and member management. When an intermediate node receives QRREQ from a source, it stores the source ID and sequence number in the cache to detect any duplicate message. It rebroadcasts the QRREQ and the routing table is updated.

4.5 Energy-Efficient Multicast Routing Protocols

The network lifetime is a key design factor of MANETs. To prolong the lifetime of MANETs, one is forced to attain the trade-off of minimizing the energy consumption and load balancing. In MANETs, energy waste resulting from retransmission due to high frame error rate (FER) of a wireless channel is significant.

4.5.1 Metrics for Energy-Efficient Multicast

1. Minimum energy constraint: In routing protocols, we need to minimize energy consumed by reducing the energy consumption through all intermediate nodes through which the packet passes. For example, from n to n_K, the packets are passing through intermediate nodes, where n is the source and n_1, n_2,

n_3...are intermediate nodes. Then, the energy consumed for all transmissions for packet j is $e_j = (n_i)$, where i is from 1 to k; the goal is to minimize the e_j.

2. Maximize time to network partition: As soon as one node in the path dies, the network is said to be partitioned, and the power consumed for each of these networks will be more. Therefore, the aim is to maximize the partition time.
3. Minimize the variance in node power levels: The idea is to treat all nodes as important nodes and balance the node energy level equally.
4. Minimize cost per packet: Choose the path into nodes as intermediate nodes, which have enough energy to reduce the cost.

Energy-efficient multicast routing protocols have the following unique characteristics:

1. Energy in wireless nodes is crucial because of the limited capacity of the battery.
2. Since nodes can move in a random way, there is frequent path failure.
3. Wireless channels have limited and more variable bandwidth compared to wired networks.

4.5.2 EEMRP: Energy-Efficient Multicast Routing Protocol

In this protocol it is assumed that the routing forwarding decision should be based on a node's energy level.

Measurement of time and energy. The following formula used to find the energy level in each node.

$$\text{Energy (E)} = \text{power} \times \text{time} \qquad (4.1)$$

That is, when a node is transmitting or receiving a packet, the energy consumption is directly proportional to transmitting or receiving power and the transmitted time.

The time is calculated as

$$\text{Time} = 8 \times \text{packet size/bandwidth} \qquad (4.2)$$

Substituting Equation 4.2 in Equation 4.1,

$$E_{tx} = P_{tx} \times 8 \times \text{packet size/bandwidth} \tag{4.3}$$

$$E_{rx} = P_{rx} \times 8 \times \text{packet size/bandwidth} \tag{4.4}$$

where E_{tx} and E_{rx} are energy consumed when the packet is transmitted and received, respectively. P_{tx} and P_{rx} are power consumed when the packet is transmitted and received, respectively. The energy consumed when nodes are forwarding a packet is equal to the sum of transmitting and receiving the packet:

$$E_t = E_{tx} + E_{rx} \tag{4.5}$$

When a node participates in forwarding a packet, the net energy is calculated as

$$\text{Energy} = E - E_t \tag{4.6}$$

When a node does not participate in forwarding a packet, the net energy is calculated as

$$\text{Energy} = E - E_s \tag{4.7}$$

where E_s is sleeping node energy. When a node does not participate in forwarding a packet, the net power is calculated as

$$\text{Power } (P) = \text{power} - \text{battery sleep power} \tag{4.8}$$

In this protocol, the energy of each node is calculated and finds the optimal route based on the energy information that is available in the node cache. The node satisfies the threshold level chosen for packet transmission. Experiments show that 70% of energy can be saved efficiently if it is implemented for a multicasting environment. When the number of nodes is increased, energy consumption is linear.

4.6 Location-Based Multicast Routing Protocols

In conventional multicasting algorithms, a collection of hosts that register to a particular group is considered a multicast group (i.e., if a host wants to receive a multicast message, first it has to join a particular group). When any hosts want to send a message to such a group, they simply multicast it to the address of that group. All the group members then receive the message.

A geocast is delivered to the set of nodes within a specified geographical area; unlike other multicast schemes, here the multicast group contains the set of nodes within a specified area. The set of nodes in a multicast region is known as a location-based multicast group and may be used for sending a message that is of interest to everyone in the specified area.

Two approaches are used to implement a location-based multicast. In a multicast tree, all nodes belonging to a multicast region belong to the multicast tree. The tree has to be updated whenever the nodes enter or leave the multicast region. In the second approach, a multicast tree is not maintained, but it uses a flooding scheme. In general, location-based multicast utilizes location information to reduce the multicast overhead. Location information is provided through a global positioning system (GPS), even though GPS does not provide accuracy up to 100%. Here, it is assumed that the error is not there.

4.6.1 Preliminaries

Multicast region and forwarding zone. Assume that a node, S, wants to multicast a message to all nodes that are currently located within a certain geographical area. The node S multicasts a data packet at time t_0, and the nodes X, Y, and Z are located within that multicast region. All three members are expected to receive the multicast data packet sent by node S. Accuracy is calculated by the ratio of number of group members that actually receive the multicast data packet and number of group members that were in the multicast region.

Forwarding zone. Node S defines "forwarding" zone (implicitly or explicitly) for the multicast data packet. A node forwards a multicast data packet only if it belongs to forwarding zone.

Location-based scheme 1. Sender S explicitly specifies the forwarding zone in its multicast data packet.

Location-based scheme 2. The forwarding zone will not be specified explicitly, the node S includes three pieces of information with its multicast data packet:

- Multicast region specification
- Location of the geometrical center (Xc,Yc) of the multicast region, distance of any node z from (Xc,Yc).

- Coordinates of sender $S(Xs,Ys)$

When a node I receives the multicast packet from node S, I determines if it belongs to a multicast region. If node I is in a multicast region, it accepts the multicast packet. Then it calculates its distance from location (Xc,Yc), denoted as $DIST_i$.

For some parameter ∂, if $DIST_s + \partial \geq DIST_i$, then node I forwards the packet to its neighbors. Before forwarding, node I replaces the (X_s, Y_s) as (X_i, Y_i).

Otherwise, if $DIST_s + \partial < DIST_i$, node I sees whether S is within a multicast region. If it is, then it will forward the packet; otherwise, it will discard it.

4.7 Summary

Any multicast routing protocol in MANETs tries to overcome some difficult problems, which can be categorized under basic issues or considerations. All protocols have their own advantages and disadvantages. One constructs multicast trees to reduce end-to-end latency. Multicast tree-based routing protocols are efficient and satisfy scalability issues; however, they have several drawbacks in ad hoc wireless networks due to the mobile nature of nodes that participate during multicast sessions. The mesh-based protocols provide more robustness against mobility and save the large size of control overhead used in tree maintenance. Most protocols of this type rely on frequent broadcasting, which may lead to a scalability problem when the number of sources increases. Hybrid multicast provides protocols, which are tree based as well as mesh based, and gives the advantages of both types. It is really difficult to design a multicast routing protocol considering all these issues.

Problems

4.1 Is hop length always the best metric for choosing paths? In an ad hoc network with a number of nodes, each differing in mobility, load generation characteristics, interference level, and so forth, what other metrics are possible?

4.2   Link-level broadcast capability is assumed in many of the multicast routing protocols. Are such broadcasts reliable? Give some techniques that could be used to improve the reliability of broadcasts.

4.3   What are the two basic approaches for maintenance of the multicast tree in bandwidth-efficient multicast protocol (BEMRP)? Which of the two performs better? Why?

4.4   What are the two different topology maintenance approaches? Which of the two approaches is better when the topology is highly dynamic? Give reasons for your choice.

4.5   How is the node energy level calculated in the EEMRP protocol?

Reference

1. Vijay, D., Deepinder, S., A multicast protocol for mobile ad hoc networks. 2001. *IEEE Conference on Communication* 3: 886–891.

Bibliography

Ballardie, A. 1997. Core based trees (CBT version 2) multicast routing. Internet Request for Comment 2.

Basagni, S., I. Chlamtac, V. R. Syrotiuk, and B. A. Woodward. 1998. A distance routing elect algorithm for mobility (dream). *Proceedings of IEEE/ACM MOBICOM'98*, pp. 76–84, October 1998.

Bommaiah, E., M. Liu, A. McAuley, and R. Talpade. AMRoute: Ad hoc multicast routing protocol. IETF MANET (draft-talpade-manet-amroute-00.txt), August 1998.

Broch, J., D. A. Maltz, D. B. Johnson, Y. C. Hu, and J. Jetcheva. 1998. A performance comparison of multihop wireless ad hoc network routing protocols. *Proceedings of ACM/IEEE MOBICOM'98*, pp. 85–97.

CMU Monarch Project. Mobility Extensions to ns-2, 1999. Available from http://www.monarch.cs.cmu.edu/

Das, S. R., R. Castaneda, J. Yan, and R. Sengupta. 1998. Comparative performance evaluation of routing protocols for mobile, ad hoc networks. *Proceedings of IEEE IC3N'98*, pp. 153–161.

Garcia-Luna-Aceves, J. J., and E. L. Madruga. 1999. A multicast routing protocol for ad hoc networks. *Proceedings of IEEE INFOCOM'99*, New York, pp. 784–792.

———. 1999. The core-assisted mesh protocol. *IEEE Journal on Selected Areas in Communication* 17 (8): 1380–1394.

Gerla, M., and S. J. Lee. On-demand multicast routing protocol for mobile ad-hoc networks. Available from http://www.cs.ucla.edu/NRL/wireless/

Jacquet, P., P. Minet, A. Laouiti, L. Viennot, T. Clausen, and C. Adjih. 202. Multicast optimized link state routing. IETF MANET draft-ietf-manet-olsr-molsr-01.txt

Joa-Ng, M., and L. T. Lu. 1999. A peer-to-peer zone based two-level link state routing for mobile ad hoc networks. *IEEE Journal in Selected Areas in Communications* (special issue on wireless ad hoc networks) 17 (8): 1415–1425.

Jogendra Kumar, G. B. Application-independent based multicast routing protocols in mobile ad hoc network (MANET). Pant Engineering College, Pauri Garhwal, Uttarakhand, India.

Johansson, P., T. Larsson, N. Hedman, B. Mielczarek, and M. Degermark. 1999. Scenario-based performance analysis of routing protocols for mobile ad-hoc networks. *Proceedings of IEEE/ACM MOBICOM'99*, pp. 195–206.

Ko, Y. B., and N. H. Vaidya. 1998. Location-aided routing in mobile ad hoc networks. *Proceedings of IEEE/ACM MOBICOM'98*, pp. 66–75.

Lee, S., J. M. Gerla, and C. C. Chiang. 1999. On-demand multicast routing protocol. *Proceedings of IEEE WCNC'99*, New Orleans, pp. 1298–1302.

Lee, S. J., M. Gerla, and C. K. Toh. 1999. A simulation study of table-driven and on-demand routing protocols for mobile ad-hoc networks. *IEEE Network* 13 (4): 48–54.

Lee, S. J., W. Su, and M. Gerla. 1999. Ad hoc wireless multicast with mobility prediction. *Proceedings of IEEE* ICCCN99, Boston, MA, Oct.

Lee, S. J., W. Su, J. Hsu, M. Gerla, and R. Bagrodia. 2000. A performance comparison study of ad hoc wireless multicast protocols. *Proceedings of the IEEE Infocom 2000.*

Liu, M., R. Talpade, A. McAuley, and E. Bommaiah. AMRoute: Ad hoc multicast routing protocol. Technical report, CSHCN T. R. 99-1, University of Maryland.

Royer, E. M., and C. E. Perkins. 1999. Multicast operation of the ad hoc on demand distance vector routing protocol. ACM MOBICOM.

Saghir, M., Wan, T-C., and Budiarto, R. 2006. Multicast routing with quality of service in mobile ad hoc networks. *Proceedings of the International Conference on Computer and Communication Engineering*, ICCCE'06 I:9–11 May 2006, Kuala Lumpur, Malaysia.

Siva Rammurty, C., and B. S. Manoj. 2008. Ad hoc wireless networks architectures and protocols, Pearson Education India, 878 pp.

Viswanath, K., Obraczka, K., and Gene Tsudik, G. Exploring mesh- and tree-based multicast routing protocols for MANETs. University of California, Santa Cruz Computer Engineering Department.

Wu, C. W., and Y. C. Tay. 1999. AMRIS: A multicast protocol for ad hoc wireless networks. *Proceedings of IEEE MILCOM'99*, Atlantic City.

5

TRANSPORT PROTOCOLS

5.1 Introduction

The transport layer acts as a liaison between the upper layer protocols and the lower layer protocols. To make this separation possible, the transport layer is independent of the physical network. The objectives of transport layer protocols include end-to end delivery of entire message, addressing, reliable delivery, flow control, and multiplexing.

Two distinct transport-layer protocols are UDP and TCP. The **user datagram protocol** (**UDP**) provides an unreliable, connectionless service to the invoking application. The second of these protocols, **TCP** (**transport control protocol**), on the other hand, offers several additional services to applications. First and foremost, it provides **reliable data transfer.** Using flow control, sequence numbers, acknowledgments, and timers, TCP's guarantee of reliable data transfer ensures that data are delivered from sending process to receiving process, correctly and in order. TCP thus converts IP's unreliable service between end systems into a reliable data transport service between processes.

TCP also uses **congestion control.** In principle, TCP permits TCP connections traversing a congested network link to share that link's bandwidth equally. This is done by regulating the rate at which the sending-side TCPs can send traffic into the network. UDP traffic, on the other hand, is unregulated. An application using UDP transport can send traffic at any rate it pleases, for as long as it pleases.

The conventional transport layer protocols that are used in wired networks are inadequate for ad hoc wireless networks because of inherent problems such as mobility, multihop route failure, hidden-and exposed-station problems, etc. that are always associated with the ad hoc networks. For example, wireless links suffer from high link error rate and TCP might interpret the packet loss caused by link error as congestion. Because of the mobility of the nodes, frequent failure and change in the route are quite certain. This route failure in turn disturbs the current TCP control mechanisms.

In response to these challenges, numerous modifications and optimizations have been proposed over the last few years to improve the performance of TCP in ad hoc networks.

The first half of this chapter discusses TCP challenges and design issues in ad hoc networks, TCP performance over mobile ad hoc networks (MANETs), and other performance. The second half focuses on transport layer protocols for ad hoc wireless networks.

5.2 TCP's Challenges and Design Issues in Ad Hoc Networks

5.2.1 Challenges

This section discusses how TCP performance degrades in ad hoc networks. This is because TCP has to face new challengers for many reasons, such as lossy channels at the physical layer, excessive contention and unfair access at the MAC (medium access control) layer, path asymmetry, network portion, failure in routes, and power constraints.

- *Lossy channels.* In ad hoc networks, wireless channels can become unavailable for several reasons.
- *Signal attenuation.* This is due to decrease in the intensity of the electromagnetic energy at the receiver because of the long distance between it and the source of transmission.
- *Doppler shift.* The relative velocity of the transmitter and the receiver is the main reason for a Doppler shift. The effect of the shift is always undesirable as there is a shift in frequency of the arriving signal, which in turn complicates the reception of the signal.
- *Multipath fading.* The radiated electromagnetic wave is always subjected to reflection and diffraction because of surrounding objects and obstacles. This causes the signal to travel over multiple paths from transmitter to receiver. Multipath propagation can further result in variations in the amplitude, phase, and geographical angle of the signal receiver at a receiver.

Fortunately, the reliable link layer protocols effectively increase the successful transmission ratio in wireless channels and mitigate the impact of lossy wireless channels on TCP performance. Of course, this done by implementing techniques: automatic repeat request (ARQ) and forward error correction (FEC).

It may be noted that the packets that are transmitted over a fading channel may cause malfunctioning of the routing protocol. Application of DSDV (destination sequence distance vector) and AODV (ad hoc on-demand distance vector routing) protocols in a real network would not provide a stable multihop route because of multipath fading behavior of the channel.

5.2.1.1 Excessive Contention and Unfair Access at MAC Layer In ad hoc networks, contention-based MAC protocols such as IEEE 802.11, where the neighboring modes contend for the shared wireless channels before transmitting, have been widely deployed. There are three important problems: the hidden terminal, the exposed terminal, and channel capture.

5.2.1.1.1 Hidden and Exposed Station Problems We referred to hidden- and exposed-station problems in the previous section. It is time now to discuss these problems and their effects.

Hidden-station problem. Figure 5.1 shows examples of the hidden-station problem. Station B has a transmission range, shown by the left oval (sphere in space); every station in this range can hear any signal transmitted by station B. Station C has a transmission range shown by the right oval (sphere in space); every station located in this range can hear any signal transmitted by C. Station C is outside the transmission range of B; likewise, station B is outside the transmission

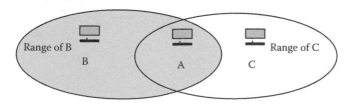

Figure 5.1 Hidden-node problem. B and C are hidden from each other with reference to A.

range of C. Station A, however, is in the area covered by both B and C; it can hear any signal transmitted by B or C.

Assume that station B is sending data to station A. In the middle of this transmission, station C also has data to send to station A. However, station C is out of B's range and transmissions from B cannot reach C; therefore, C thinks the medium is free.

Station C sends its data to A, which results in a collision at A because this station is receiving data from both B and C. In this case, we say that stations B and C are hidden from each other with respect to A. Hidden stations can reduce the capacity of the network because of the possibility of collision.

The solution to the hidden-station problem is the use of the handshake frames like RTS (request to send) and CTS (clear to send).

Exposed-station problem. Now consider a simulation that is the inverse of the previous one: the exposed-station problem. In this problem a station refrains from using a channel when it is in fact available. In Figure 5.2, station A is transmitting to station B, and station C has some data to send to station D that can be sent without interfering with the transmission from A to B. However, station C is exposed to transmission from A; it hears what A is sending and thus refrains from sending. In other words, C is too conservative and wastes the capacity of the channel.

The handshaking message of RTS and CTS cannot help in this case. Station C hears the RTS from A, but does not hear the CTS from B. After hearing the RTS from A, station C can wait for a time so that the CTS from B reaches A; it then sends an RTS to D to show that it needs to communicate with D. Both stations B and

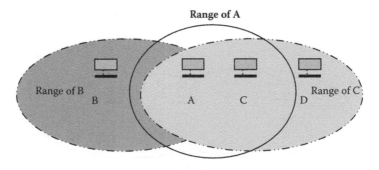

Figure 5.2   Exposed-node problem. C is exposed to transmission from A to B.

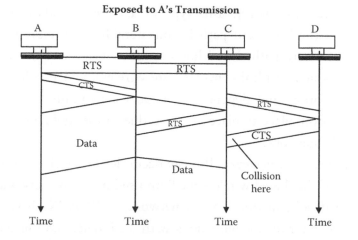

Figure 5.3 Handshaking messages during exposed-node problem.

A may hear this RTS, but station A is in the sending state, not the receiving state. Station B, however, responds with CTS. The problem is here. If station A has started sending its data, station C cannot hear the CTS from station D because of the collision; it cannot send its data to D and remains exposed until A finishes sending its data, as Figure 5.3 shows.

Channel capture. In addition, the aggressive behavior of TCP and its poor interaction with the MAC layer further exacerbate the unfairness situation. In extreme cases, a few TCP flows capture the channel, and other TCP flows cannot access it for some amount of time, leading to similar false link failure.

Path asymmetry. Path asymmetry in TCP-based wireless ad hoc networks can be classified as the following types:

- *Bandwidth asymmetry.* This type of asymmetry is found in satellite networks in which forward and backward data flow in different paths with different speeds. In ad hoc networks, this can happen as well, since not necessarily all nodes have the same interface speed. So, even if a common path is used in both directions of a given flow, they do not necessarily have the same bandwidth. In addition, as the routing protocols can assign different paths for forward and backward traffic, asymmetry is certain in wireless ad hoc networks.

- *Loss rate asymmetry.* This type of asymmetry takes place when the backward path is considerably more lossy than the forward path. In ad hoc networks, this can be a serious issue as all links involved are wireless, which is highly error prone and dependent on local constraints that can vary from place to place and also due to the mobile nature of the network.

- *Media access asymmetry.* This can occur due to the characteristics of the shared wireless medium used in ad hoc networks. Specifically, in this kind of network, TCP ACKs have to contend for the medium along with TCP data, and this may cause excessive delay as well as discarding on TCP ACKs.

- *Route asymmetry.* Unlike the preceding three forms of asymmetry, where forward and backward paths can be the same, route asymmetry implies distinct paths in both directions. Route asymmetry is associated with the possibility of different transmission ranges for the nodes in this scenario. In fact, the transmission range of each node depends on its instantaneous battery power level, which, in most cases, is likely to vary over time. The inconvenience with different transmission ranges is that it can lead to conditions in which the forward data follow a considerably shorter path than the backward data (TCP ACK). Due to lack of power in one (or more) of the communications with the destination, it has to communicate through a multihop connection. However, multihop connections are prone to be low-throughput effective. Consequently, the TCP ACKs may face considerable disruption. Furthermore, mobility and variation in the battery power level make the problem even worse since they may cause frequent route change.

To summarize, all these types of asymmetry may ultimately result in damaging the forward throughput and lead to inaccuracy on RTT estimation. In order to improve TCP performance, three techniques have been proposed that may be useful in ad hoc networks:

1. TCP header compression is based on the fact that most of the field of TCP header compression has been proposed to reduce the size of TCP ACK packets in the backward path.
2. ACK filtering reduces the number of TCP ACKS transmitted in the backward path. This scheme takes advantage of the fact that the ACK packets are cumulative.
3. ACK congestion control causes the receiver to control the congestion on the backward path.

Network partition. Mobile terminals in ad hoc networks can be regarded as simple graphs in which mobile terminals are the "vertices" and a successful transmission between two terminals is an edge. Whenever there is a disconnection in such a graph because of random movement of mobile nodes in an ad hoc wireless network, this can lead to network partitions. Energy-contained operation of nodes is also another cause for network partition; as is evident from Figure 5.4, when node D moves away from node C, this results in a partition of the network. The TCP agent of node A cannot receive the TCP ACK transmitted by F.

If the partition persists for a duration greater than retransmission time-out (RTO) of the node A, the TCP agent stores the exponential back-off, which consists of doubling the RTO whenever the time-out expires. TCP has no information about the time of network reconnection. This lack of information may lead to long idle periods during which the network is connected but TCP is in the back-off state.

Routing failures. Node mobility and contention on the wireless channel are the two important causes of routing failure. The reestablishment of route and its duration after route failure in ad hoc networks depends on facts like mobility pattern of mobile nodes, traffic characteristics, and the underlying routing protocol. As already discussed

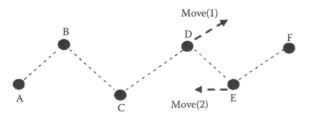

Figure 5.4 Example for network partition.

in the previous section, TCP senders do not have information on the route reestablishment event, throughput a session delay will designed because of large idle time.

Also, if the new route established is not comparable with old route, then TCP will come across a large fluctuation in routed trip time (RTT).

In ad hoc networks, routing protocols that depend on broadcast Hello messages to find neighbor nodes may suffer from the "communication gray zones" problem. In such zones, data messages cannot be exchanged even if a neighbor's node are reachable as is indicated by Hello messages and control framer. This is how routing protocols will experience routing failures.

Lundgren et al. [1] have conducted experiments and have subsequently concluded that the origin of this problem is heterogeneous transmission routes, absence of ACKs for broadcast packets, small packet size of Hello messages, and fluctuations of wireless links.

Power constraints. Mobile nodes are battery powered devices and, because of mobility, nodes have limited power supply due to which processing power is limited. This is a major issue in ad hoc wireless networks, as each node acts as a router and as an end station; obviously, additional energy is required to forward and reradiate the packets.

It is the responsibility of TCP to use this scarce power source in an "efficient" manner—that is, minimizing the number of unwanted retransmissions at the transport layer as well as at the link layer. In ad hoc wireless networks, there are two correlated power problems: power saving and power control.

Power saving strategies have been investigated at several levels of mobile nodes, including power layer transmission, operation systems, and applications. Power control is achieved by adjusting the transmission power of mobile devices. Power control can be jointly used with routing and transporting agents to improve the performance of ad hoc networks. Power constraint communications reveal also the problem of cooperation between nodes, as nodes may not participate in routing and forwarding procedures in order to save battery power.

5.2.2 Design Goals

- The transport layer protocol should maximize the throughput per connection.

- The transport layer protocol should provide fairness across competing flows.
- The transport layer protocol should have reduced e-connection setup and connection maintenance overheads. The protocol must facilitate scalability in large networks by reducing the requirements for setting up and maintaining the connections
- The protocol should provide means and measures for congestion control and flow control in the ad hoc wireless network.
- The protocol should be able to offer both reliable and unreliable connections.
- The protocol should be able to adapt to the mobility and change in topology of the ad hoc wireless network.
- One of the important resources must be used efficiently.
- The protocol should be aware of limitations and resource constraints.
- Like any other layers in a network, this protocol should make use of information from the lower layer.
- The protocol should offer a well-defined cross-layer interaction framework.
- This protocol should also maintain end-to-end semantics.

5.3 TCP Performance over MANETs

5.3.1 TCP Performance

In this section the TCP performance over mobile ad hoc networks is discussed. The effect and impact of mobility on TCP throughput in MANETs has been investigated by Monks, Sinha, and Bharghavan [2]. According to their simulation report, nodes move according to the random-way point model with pause time of 0 s. The speed of the node was uniformly distributed in [0.9v–1.1v] for some mean speed v.

By applying the dynamic source routing (DSR) algorithm at the routing layer, the author reports that when the speed increases from 2 to 10 m/s, the throughput drops sharply. However, it is found that there is only a slight drop in throughput when the mean speed is increased from 10 to 30 m/s. Also, the authors remark that, for a given mean speed, certain mobility patterns result in throughput close to zero, even though the other mobility patterns are able to achieve high throughput.

After careful analysis of the simulation trace of patterns with throughput, the authors found that the TCP senders' routing protocol is unable to recognize and flush out stale routes from its cache, which in turn leads to repeated failures in routing and TCP retransmission time-outs. They found that, most of the time, the TCP sender and receiver are close to each other.

From the nature of the mobility patterns, the authors observed that, as the sender nodes and receiver nodes move closer to each other, DSR can maintain a valid route by shortening the existing route before a routing failure occurs. But, as sender and receiver move away from each other, DSR waits until a failure occurs to lengthen a route. The route failure induces up to a TCP-window's worth of packet losses and the subsequent route discovery process may result in repeated TCP time-outs, which are called "serial time-outs."

Losses that are induced by mobility of nodes may cause TCP invocation of congestion control that deteriorates the TCP throughput. So, in order to prevent such TCP invocation, the authors suggest using the explicit link failure notification technique.

One of the main problems that TCP has over MANETs is that "TCP treats loss induced by route failure as signs of network congestion." Anantharaman et al. [3] identify a number of factors that contribute to the degradation of TCP throughput when nodes are mobile. The two important factors that are responsible for such degradation of TCP performance are

1. MAC failure detection latency
2. Route recomputation latency

MAC failure detection latency is defined as the amount of time spent before the MAC concludes a link failure. The authors found that in the case of the IEEE 802.11 protocol, this latency is small and independent of the speed of the moving nodes, when the load is light (one TCP connection). However, in the case of high loads, they observed that the value of this latency is magnified and becomes a function of the node's speed.

Route computation latency is defined as the time taken to recompute the route after a link failure. They found that, as for MAC failure detection latency, the route computation latency increases with the load and becomes a function of the node's speed in the high load case. Also, the authors identify another problem, called **MAC packet arrival,** that is related to routing protocols. In fact, when a link failure is detected, the link failure is sent to the routing agent of the packet that triggered the detection. If other sources are using the same link in the path to their destinations, the node that detects the link failure has to wait till it receives a packet from these sources before they are informed of the failure. This also contributes to the delay after a source realizes that a path is broken.

Dyer and Boppana [4] report simulation results on the performance of TCP Reno over three different routing protocols (ad hoc on demand vector [5], dynamic source routing [6], and ADV [7]). It is found that ADV performs well under a variety of mobility patterns and topologies. Furthermore, they propose a heuristic technique called fixed RTO to improve the performance of on-demand routing protocols (AODV and DSR). According to this technique, the TCP's performance degrades when the multipath routing protocol SMR [8] is used. Multipath routing affects TCP by two factors: the inaccuracy of the average RTT measurement that leads to more premature time-outs and the out-of-order packet delivery via different paths, which triggers duplicated ACK, which in turn triggers TCP congestion control.

5.3.2 Other Problems

5.3.2.1 State Route Problem   The mobility nodes and change in topology may tend to change in routes, as a result of which there is a need for updating routes as soon as possible. The TCP sender is very slow change in topologies, the router from its cache, resulting in very frequent failure in routing. And intermediate nodes may reply to route requests with these state routes in their cache. This complicates the problem of routing, and the problem gets still worse when neighbor nodes overhear the state routed in replies. Therefore, state routes are spread throughout the network, causing further route failure in the network. The ultimate effect is an adverse effect on TCP performance. It can be solved by adjusting the route cache timing depending on route failure route.

5.3.2.2 MAC Layer Rate Adaptation Problem   The MAC layer adaptation algorithm is supposed to increase the throughput when there is a high channel rate. But a poor rate adaptation algorithm could decrease the throughput. The multiplicative increase–multiplicative decrease rate algorithm causes the periodic retransmissions of TCP packets. This further causes network trashing on wireless local networks.

Because of MAC layer retransmission, there will be a waste of channel resources, so there is a need for a better rate adaptation algorithm. There are many problems that cause TCP performance degradation in mobile ad hoc networks, among which the following are more important:

- The ability of TCP to distinguish between packet losses due to congestion route failure
- The TCP suffering from frequent route failure
- The contention on the channel
- The unfairness problem of the TCP

5.4 Ad Hoc Transport Protocols

5.4.1 Split Approaches

The fairness and throughput of TCP suffer when it is used in mobile area networks; that is, as length of the path increases, the overall degradation of throughput also increases. The short connections (i.e., in terms of path length) enjoy an unfair advantage over long connections, which means the short connections generally obtain higher throughput as compared with longer connections. This can also lead to unfairness among TCP sessions, where one session may obtain much higher throughput than other sessions.

This unfairness problem is further worsened by the use of the MAC protocol, which is commonly used in ad hoc networks. This MAC protocol is described in the IEEE 802.11 standard.

One specific problem that is induced by the explanations of back-off mechanisms of the IEEE 802.11 MAC protocol is the "channel capture effect." Because of this effect, the most data-intense connection dominates the multiple access wireless channels. If there are multiple data-intense connections, the first connection "captures" the channel until it has transported all of its data to the destinations.

This leads to unfairness to the connection that begins later or further away from the point of contention. Hence, once again, the connections with a large number of hops are at a disadvantage.

The spilt-TCP approach provides a unique solution to this problem; the scheme splits the transport layer objectives into congestion control and reliable packet delivery.

Congestion control is a local phenomenon due to high contention and high traffic load in local regions. In the mobile ad hoc wireless network environment, this demands a local solution. At the same time, reliable packet delivery is an end-to-end acknowledgment.

In addition to splitting the transport layer functionalities, split-TCP splits long TCP connections into shorter localized segments or zones. This is done in order to improve the performance in terms of fairness.

To substantiate this idea, the split-TCP scheme uses a number of selected intermediate nodes between these localized segments known as **proxy nodes.** Using this scheme, if a packet needs to be transmitted, the proxy node receives the TCP packets; when it intercepts TCP packets, it reads the content of packets, buffers them in its local buffer, and send on acknowledgment to the source (or previous proxy node). This acknowledgment is known as local acknowledgment (LACK) and the proxy node takes over the responsibility of delivering the packets further, at an appropriate rate, to the next local segments.

Upon the receipt of a LACK (from the next proxy or from the final destination), a proxy will purge the packet from its buffer. The forwarded packet could possibly be intercepted again by another proxy and so on. In this scheme, there is no change in the end-to-end acknowledgment system of TCP, meaning that the source will not clear a packet from its buffer unless it is acknowledged by a cumulative ACK from the destination. However, it may be noted that the overhead incurred in including infrequent end-to-end ACKs in addition to the LACKs is extremely small and can be considered to be acceptable, given the advantages of split TCP.

In Figure 5.5, node S initiates a TCP session to the node D. Nodes p1 and p2 are chosen as proxy nodes. The number of proxy nodes in a TCP session is determined by the length of the path between source and destination nodes.

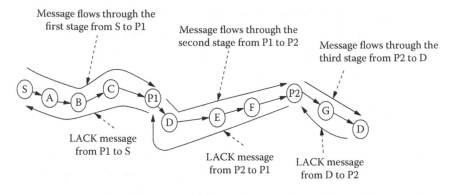

Figure 5.5 TCP with proxies p1 and p2.

The proxy node p1, upon receipt of each TCP from the source node S, acknowledges with a LACK packet and buffers the received packets. This buffered packet is forwarded to the next proxy (in this case, node p2) at a rate proportional to the rate of arrival of LACKs from the next proxy node or destination.

The source keeps transmitting about purges a packet from its buffer only upon receipt of an end-to-end ACK for that packet from the destination. (Note that this might be indicated in a cumulative ACK for a plurality of packets.)

This scheme essentially splits the transport layer functionalities into those of congestion control and end-to-end reliability. Correspondingly, the transmission control window at the TCP ender is split into two windows: the congestion window and the end-to-end window. The congestion window will always be a subwindow of the end-to-end window. While the congestion window changes in accordance with the rate of arrival of LACKs for the next proxy, the end-to-end window will change in accordance with the rate of arrival of end-to-end ACKs from the destination.

5.4.2 End-to-End Approach

The end-to-end approach mainly addresses the problem of a TCP's misinterpretation of packet losses due to route failure due to network congestion in mobile ad hoc networks.

5.4.2.1 TCP Feedback (TCP-F) In order to improve the performance of an ad hoc wireless network, traditional TCP has been modified. TCP feedback employs a feedback-based approach. The TCP sender gets TCP-F, which relies on the support of a reliable link layer and routing protocol. It is the responsibility of the routing protocol to repair the broken links within a desirable time period.

TCP-F intends to minimize the throughput degradation resulting from rapid change in mobility of topology due to the mobility of mobile hosts. The frequent change in topology of mobile nodes leads to packet losses. TCP misinterprets such losses as congestion and invokes congestion control, leading to unnecessary retransmission and degradation of throughput. In feedback-based TCP, whenever an intermediate node detects the link break, the intermediate nodes sends a route failure notification (RFN) packet toward the sender. This intermediate node maintains the information about all the RFN packets it has originated so far and updates its routing table accordingly.

When a TCP sender receives the RFN, the sender goes into the snooze state. As soon as it enters the snooze state, the sender stops sending any more packets to the destination, freezes all of its timers, freezes its congestion window, and sets up a route failure timer, which is the time required to reestablish with the route. The route failure timer that is thus initiated is dependent on network size, network topology, and network protocol. When this timer expires, the sender changes its state from the snooze to the achieve state and, of course, the sender node receives the information regarding reestablishment of the route from an intermediate node through a route re-establishment notification (RRN) packet.

As soon as a sender gets the RRN packet, it transmits all the packets in its buffer, assuming that the network is back to its active or connected state.

Figure 5.6 clearly explains the operation of the TCP-F protocol. This figure shows a TCP session set up between nodes S and D. When the intermediate link between node IN_2 and D fails, the intermediate node originates an RFN packet and sends that packet in the reverse direction toward the source. After receiving the RFN packet, the source node enters a snooze state. It remains in its snooze state till it receives another packet called the RRN packet. This notifies about reestablishment of the link. Advantages include the following:

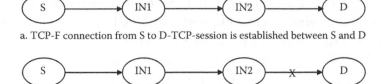

a. TCP-F connection from S to D-TCP-session is established between S and D

b. Link IN2-D breaks and originate RFN-sender enters to snooze-state

c. Link IN2-D reestablished. RRN have been sent to the sender

Figure 5.6   Operation of TCP-F protocol.

- Simple feedback brings a good solution to minimize the problem due to frequent failure in the links.
- It is a good congestion control mechanism.

The following is a disadvantage:

- Implementation of feedback-based TCP requires modification to existing TCP libraries.

5.4.2.2 TCP-ELFN   TCP-ELFN is similar to TCP-F; however, in contrast to TCP-F, the evolution of the proposal is based on a real interaction between TCP and the routing protocol. Such interaction is required to inform the TCP agent about link and route failures so that it can avoid responding to the failure as if to congestion.

In TCP-ELFN, the explicit link failure notification (ELFN) packet is originated by the intermediate node detecting path breaks upon detection of a link failure to the sender. To implement an ELFN message, the route failure message of DSR is modified to carry a payload similar to a "host unreachable" ICMP (Internet control message protocol) message.

Upon receiving an ELFN, the TCP sender disables its retransmission timers and enters into a "standby" state by freezing the regular transmission of packets until the connection is reestablished. Then the transmission is resumed. During this node of standby, the TCP sender periodically sends a small packet to probe the network to see if a route has been established. Upon reception of an acknowledgment packet for the probe packets, it comes out of standby mode and restores the retransmission timer and continues to perform regular transmission of packets.

5.4.2.3 Ad Hoc-TCP Similarly to TCP-F and TCP-ELFN, ATCP (ad hoc transmission control protocol) utilizes a network layer feedback mechanism with which the TCP sender can come to know the status of a network path through which TCP packets are propagated. The TCP sender can be put into a persistence state, congestion control state, or retransmit state, depending upon the feedback information that it gets from intermediate nodes.

As soon as a network partition is noticed by an intermediate node, the TCP sender enters the persistence state, where it avoids unnecessary retransmissions. During this state, the TCP sender sets TCP's congestion window size to one in order to ensure that TCP does not continue to use old congestion window values. This forces TCP to probe the correct value of the congestion window to be used for the new route. If an intermediate node encounters packet loss due to error, then the ATCP immediately retransmits it without invoking the congestion control algorithm.

In order to be compatible with widely deployed TCP-based networks, ATCP provides this feature without modifying the traditional TCP. ATCP is shown in Figure 5.7 and it is implemented as a thin layer between network layer and transport layer without any changes in the existing TCP protocol. The import function of the ATCP layer keeps track of the packet sent and received by the TCP sender, the state of the network, and the state of the TCP sender.

As shown in the ATCP state diagram, the four possible states are normal, congested, loss, and disconnected. ATCP at the sender is in the normal state. In this state, ATCP does nothing and it remains invisible. In a lossy channel, it is likely that some packets are lost or may arrive out of order. This receiver generates duplicate ACKs. In the case of traditional TCP, upon reception of three consecutive duplicate ACKs, it transmits the offending segment and shrinks the congestion window.

But ATCP in its normal state counts the number of duplicate ACKs received for any segment. When it sees that three duplicate ACKs have been received, it does not forward the third duplicate ACK, but puts TCP in persist node and ATCP in the loss state. Hence, the TCP sender avoids invoking congestion control.

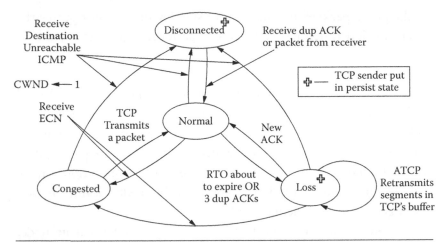

Figure 5.7   State transition diagram for ATCP at sender.

In the loss state ATCP transmits the unacknowledged segments from TCP's send buffer. When a new ACK arrives from the TCP receiver, ATCP forwards that ACK to TCP, which also removes TCP from the persist node. ATCP then returns to its normal state.

When the ATCP sender is in the loss state, receipt of an ECN message changes it to a congested state. In addition to this transition, the ATCP sender removes the TCP sender from the persist state. When the network detects congestion, the ECN flag is set in ACK and data packets.

Let us assume that a TCP receives this message when in its normal state. ATCP moves into its congested state and does nothing. It ignores any duplicate ACKs that arrive. In other words, ATCP does not interfere with TCP's normal congestion behavior. After TCP transmits a new segment, ATCP returns to its normal state. Mobility of nodes in ad hoc networks causes route failure or a transient network partition. When this happens, ATCP expects the network layer to detect these and inform the ATCP sender through an ICMP destination unreachable message.

When ATCP receives this message, it puts the TCP sender into persist mode and itself enters the disconnected state. It continues to be in the DISCONN state until it is connected and receives any data or duplicate ACKs. On the occurrence of any of these events, ATCP changes to the normal state. TCP periodically generates probe packets and this is done in order of the path. The receipt of an ICMP

DUR message in loss state or congested state causes a transition to the DISCONN state.

When ATCP puts TCP into the persist state, it sets the congestion window to one segment. This is done in order to make TCP probe for the new congestion when the new route is available.

ATCP offers two important advantages:

1. Significant improvement in TCP performance while maintaining the end-to-end semantics of TCP
2. Compatibility with traditional TCP

Its disadvantages include:

1. Dependence on network layer to detect route failure and network partitions
2. Inclusions of a thin ATCP layer to the TCP/IP protocol stack that need changes in inter foretimes

5.4.2.4 TCP-Buffering Capability and Sequencing Information (TCP-BUS) TCP-BUS is similar to TCP-F and TCP-ELFN and it uses the network feedback to detect route failure and to take react suitably to such failures. This incorporates buffering capability in mobile nodes and uses associatively based routing (ABR) as a routing scheme.

TCP-BUS makes use of some of the special messages such as localized query (LQ) and REPLY, defined as part of ABR for finding a partial path.

These control messages are modified and intended to carry TCP connection and segment information. At the source, the TCP-BUS sender transmits its segments in the same manner as general TCP, when there are no feedback messages. However, upon the detection of a path break, an intermediate node called the pivot node (PN) originates an explicit route disconnection (ERDN) feedback message. The ERDN feedback message is sent back to the TCP-BUS sender. When a source receives the ERDN feedback message, it stops sending data packets. In addition, it freezes all timer values and window sizes as in TCP-F.

Packets in transit at the intermediate from TCP-BUS sender to the PN are buffered until a new partial path from this intermediate node to the TCP-BUS receiver is formed by the pivot node. The timers for all the buffered packets at various intermediate nodes, source nodes, and pivot nodes use time at values proportional to round-trip time (RTT). This is required to avoid unnecessary retransmissions.

The nodes between TCP-BUS sender and PN can request the TCP-BUS sender to transmit any of the lost packets selectively.

Upon detection of a path break, the downstream node originates as a route notification (RN) packet to the TCP-BUS destination node, which is forwarded by all the downstream nodes in the path. This in turn invalidates the old partial path and flushes out buffered packets along that path.

The ERDN packet is sent to the TCP-BUS sender in a reliable way using an implicit acknowledgment and retransmission mechanism.

The PN node includes the sequence number of the TCP segment belonging to the flow that is currently at the head of its queue in the ERDN packet. The PN also attempts to find a new partial path to the TCP-BUS receiver. Availability of such a partial path to a destination is explicitly intimated to a TCP-BUS sender through a route successful notification (RSN) message.

TCP-BUS uses the route reconfiguration mechanism of ABR to set the partial route to the receiver node. This needs other routing protocols to be modified to support TCP-BUS.

The control messages LQ and REPLY are modified to carry the sequence number of the segment at the need of the queue buffered at PN and sequence number of the last successful segment the TCP-BUS receiver received. The LQ packet carries the sequence number of the last successful segment the TCP-BUS receiver received.

This makes the TCP-BUS receiver understand the packets lost in transition and those buffered at the intermediate nodes. This is used to avoid fast retransmission packet delivery. Upon a successful LQ-REPLY process to obtain a new route to the TCP-BUS receiver, PN informs the TCP-BUS sender of the new partial path using the ERSN packet. When the TCP-BUS sender receives an ERSN packet, it resumes the data transmission.

Since there is a chance for ERSN packet loss due to congestion in the network, it needs to be sent reliably. That TCP-BUS sender also periodically originates probe packets to check the availability of a path to the destination. Figure 5.8 shows an illustration of the propagation of ERDN and RN messages when a link between nodes 4 and 12 fails.

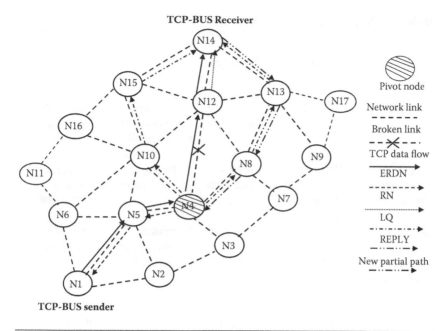

Figure 5.8 Operation of TCP-BUS.

When a TCP-BUS sender receives the ERSN message, it under-stands, for the sequence number of the last successfully received pack-ets at the destination and the sequence number of packets at the head of the queue at PN, that the packets will be delayed further and hence uses a selective acknowledgment strategy instead of fast retransmis-sion. The last packets are retransmitted by the TCP-BUS sender. During the retransmission of this last packet, the network congestion between the TCP-BUS sender and PN is handled in a way similar to that in traditional TCP.

Advantages include:

- Improved performance
- Uses of buffering, sequence numbering, and selective acknowl-edge, thus avoiding fast retransmission

Disadvantages include:

- More dependency on routing protocol and buffer at interme-diate nodes
- Performance adversely affected in the event of failure of inter-mediate nodes

Table 5.1   Comparison between Different End-to-End Approaches

	TCP-F	ELFN	ATCP	TCP-BUS
High BER packet loss	Not handled	Not handled	Handled	Not handled
Route failure (RF) detection	RFN packet freezes TCP Sender state	ELFN packet freezes TCP Sender state	ICMP "destination unreachable" freezes TCP Sender state	ERDN packet freezes TCP Sender state
Route reconstruction (RR) detection	RRN packet resumes TCP to normal state	Probing mechanism	Probing mechanism	ERSN packet resumes TCP to normal state
Packet reordering	Not handled	Not handled	Handled	Not handled
Congestion window and retransmission time-out (RTO) after RR	Old CW and RTO	Old CW and RTO	Reset for each new route	Old CW and RTO
Reliable transmission of control messages	Not handled	Not handled	Not handled	Handled
Evaluation	Emulation; no routing protocol considered	Simulation	Experimental; no routing protocol considered	Simulation

Table 5.1 gives an illustration of the comparison among the different end-to-end approaches of TCP protocols.

5.5 Summary

This chapter discussed the major challenges that a transport layer protocol faces in mobile ad hoc wireless networks. The important design goals of a transport layer protocol were listed, and the chapter also provided a classification of existing transport layer solutions. TCP is one of the important and most widely used transport layer protocols and is regarded as the backbone of today's Internet. It provides end-to-end, reliable, byte-streamed, in-order delivery of packets to nodes.

Because TCP was designed to handle problems present in traditional wired networks, many of the issues that are present in dynamic topology networks such as ad hoc wireless networks are not addressed.

This causes reduction of throughput when TCP is used in ad hoc wireless networks. It is very important to use TCP in ad hoc wireless networks as it is important in seamless communication with the Internet whenever and wherever it is available. This chapter provided a discussion on the major reasons for the degradation in the performance of traditional TCP in ad hoc wireless networks and explained a number of recently proposed solutions to improve TCP's performance.

References

1. Lundgren et al.
2. Monks, J. P., P. Sinha, and V. Bharghavan. 2000. Limitations of TCP-ELFN for ad hoc networks. Workshop on Mobile and Multimedia Communication, Marina del Rey, CA, Oct. 2000.
3. Anantharaman et al.
4. Dyer, T. D., and R. Bopanna. 2001. A comparison of TCP performance over three routing protocols for mobile ad hoc net works. *Proceedings of ACM MOBIHOC 2001,* Long Beach, CA, Oct. 2001.
5. Perkins, C., and T. Watson. 1994. Highly dynamic destination-sequenced distance-vector routing (DSDV) for mobile computers, in *Proceedings of ACM SIGCOMM,* London, UK.
6. Johnson, D., D. Maltz, and Y. Hu. 2003 (Internet draft). The dynamic source routing protocol for mobile ad hoc networks (DSR).
7. Boppana, R., and S. Konduru. 2001. An adaptive distance vector routing algorithm for mobile, ad hoc networks, in *Proceedings of IEEE INFOCOM,* Anchorage, AK. 2001.
8. Lu, S., and M. Gerla, Split multipath routing with maximally disjoint paths in ad hoc networks, in *Proceedings of IEEE ICC,* Helsinki, Finland.
9. Holland, G., and N. Vaidya. 2002. Analysis of TCP Performance over mobile ad hoc networks, *ACM Wireless Networks,* 8 (2): 275–288.
10. Ramakrishnan, K., S. Floyd, and D. Black. 2001. *The Addition of Explicit Congestion Notification* (ECN) to IP.

Bibliography

Bakre, A., and B. Badrinath. 1995. I-TCP: Indirect TCP for mobile hosts. *Proceedings of 15th International Conference on Distributed Computing Systems (ICDCS),* Vancouver, BC, Canada, May.
Bakshi, B., P. Krishna, N. H. Vaidya, and D. K. Pradhan. 1997. Improving performance of TCP over wireless networks. *Proceedings of 17th International Conference on Distributed Computing Systems (ICDCS),* Baltimore, MD, May.

Balakrishnan, H., and S. Seshan. 1995. Wireless networks. *Proceedings of ACM MOBICOM,* Berkeley, CA, Nov.

Chandran, K., S. Raghunathan, S. Venkatesan, and R. Prakash. 1998. A feedback-based scheme for improving TCP performance in ad-hoc wireless networks. *Proceedings of International Conference on Distributed Computing Systems,* Amsterdam, May, pp. 472–479.

Chen, T., and M. Gerla. 1998. Global state routing: A new routing scheme for ad hoc wireless networks. *Proceedings of IEEE ICC'98,* Aug., pp. 171–175.

Chiang, C. C., H. K. Wu, W. Liu, and M. Gerla. 1997. Routing in clustered multihop mobile wireless networks with fading channel. *Proceedings of IEEE Singapore International Conference on Networks SICON'97,* April, pp. 197–212.

Corson, M. S., and J. Macker. 1999. Mobile ad hoc networking (MANET): Routing protocol performance issues and evaluation considerations. Request for Comments 2501, *IETF,* Jan.

Das, S. R., R. Castaneda, and J. Yan. 1998. Comparative performance evaluation of routing protocols for mobile ad hoc networks.

Feeney, L. M. 1999. A taxonomy for routing protocols in mobile ad hoc networks. SICS technical report T99/07, Oct. 1999 (http://citeseer.ist.psu.edu/feeney99 t axonomy.html).

Gerla, M., X. Hong, and G. Pei. 2001. Fisheye state routing protocol (FSR) for ad hoc networks. IETF draft.

Handley, M., C. Bormann, B. Adamson, and J. Macker. 2003. NACK oriented reliable multi-cast (NORM) protocol building blocks. Internet draft, RMT Working Group (draft-ietf-rmt-bb-norm-05.txt).

Henderson, T., and R. Katz. 1997. Satellite transport protocol (STP): An SSCOP-based transport protocol for datagram satellite net works. *Proceedings of 2nd Workshop on Satellite-Based Information Systems (WOSBIS),* Budapest, Hungary.

Jaquet, P., P. Muhlethaler, and A. Qayyum. 2001. Optimized link-state routing protocol. IETF draft.

Jiang, X., and T. Camp. 2002. A review of geocasting protocols for a mobile ad hoc network. Grace Hopper Celebration (GHC) (http://toilers.mines.edu/papers/pdf/Geocast-Review).

Koksal, C. E., and H. Balakrishnan. 2000. An analysis of short-term fairness in wireless media access protocols (poster). *Proceedings of ACM SIGMETRICS, Measurement and Modeling of Computer Systems,* Santa Clara, CA, pp. 118–119.

Liu, J., and S. Singh. 2001. ATCP: TCP for mobile ad hoc networks. *Communications* 19 (7): 1300–1315.

Madruga, E. L., and J. J. Garcia-Luna-Aceves. 2001. Scalable multicasting: The core-assisted mesh protocol. *Mobile Networks and Applications* 6 (2): 151–165.

Murthy, S., and J. J. Garcia-Luna-Aceves. 1996. An efficient routing protocol for wireless networks. *ACM Mobile Networks and Applications* 1:183–197.

6

QUALITY OF SERVICE

6.1 Introduction

For supporting multimedia applications, it is desirable that an ad hoc network has a provision of quality of service (QoS). However, providing the QoS in a mobile ad hoc network is a challenging task. Quality of Service (QoS) means that the network should provide some kind of guarantee or assurance about the level or grade of service provided to an application. The definition for QoS and the QoS parameter may be considered different for different applications, which purely depends upon specific requirements of an application. For example, an application that is delay sensitive may require the QoS in terms of delay guarantees. Some applications may require that the packets should flow at certain minimum bandwidth. In that case, the bandwidth will be a QoS parameter. The other application may require a guarantee that the packets are delivered from a given source to a destination reliably; then, reliability will be a parameter for QoS.

6.2 Challenges

The characteristics of an ad hoc network pose several challenges in the provision of QoS. Some of these challenges are as follows:

- *Dynamically varying network topology.* Since the nodes in an ad hoc wireless network do not have any restriction on mobility, the network topology changes dynamically. Hence the admitted QoS sessions may suffer due to frequent path breaks, thereby requiring such sessions to be re-established over new paths. The delay incurred in re-establishing a QoS session may cause some of the packets belonging to that session to miss their delay targets/deadlines, which is not acceptable for applications that have stringent QoS requirements.

- *Imprecise state information.* In most cases, the nodes in an ad hoc wireless network maintain both the link-specific state information and flow-specific state information. The link-specific state information includes bandwidth, delay, delay jitter, loss rate, error rate, stability, cost, and distance values for each link. The flow-specific information includes session ID, source address, destination address, and QoS requirements of the flow (such as maximum bandwidth requirement, minimum bandwidth requirement, maximum delay, and maximum delay jitter). The state information is inherently imprecise due to dynamic changes in network topology and channel characteristics. Hence, routing decisions may not be accurate, resulting in some of the real-time packets missing their deadlines.

- *Lack of central coordination.* Unlike wireless LANs and cellular networks, mobile ad hoc networks (MANETs) do not have central controllers to coordinate the activity of nodes. This further complicates QoS provisioning in MANETs.

- *Error-prone shared radio channel.* The radio channel is a broadcast medium by nature. During propagation through the wireless medium, the radio waves suffer from several impairments, such as attenuation, multipath propagation, and interference (from other wireless devices operating in the vicinity).

- *Hidden-terminal problem.* The hidden-terminal problem is inherent in MANETs. This problem occurs when packets originating from two or more sender nodes that are not within the direct transmission range of each other collide at a common receiver node. This necessitates retransmission of packets, which may not be acceptable for flows that have stringent QoS requirements. The RTS/CTS control packet exchange mechanism adopted in the IEEE 802.11 standard reduces the hidden-terminal problem only to a certain extent.

- *Limited resource availability.* As MANETs have limited resources such as bandwidth, battery life, storage space, and processing capability, they have to be utilized in a very efficient way. Out of these, bandwidth and battery life are considered as very critical resources, the availability of which significantly affects the performance of the QoS provisioning mechanism. Hence, efficient resource management mechanisms are required for optimal utilization of these scarce resources.

• *Insecure medium.* Security in a wireless channel is considerably less, due to the broadcast nature of the wireless medium. Hence, security is an important issue in MANETs, especially for military and tactical applications. MANETs are susceptible to attacks such as eavesdropping, spoofing, denial of service, message distortion, and impersonation. Without sophisticated security mechanisms, it is very difficult to provide secure communication guarantees.

The design choices for providing QoS support are described below.

6.2.1 Hard-State versus Soft-State Resource Reservation

In any QoS framework, QoS resource reservation is a very important component. (A QoS framework can be considered as a complete system that provides required/promised services to each user or application). It is responsible for reserving resources at all intermediate nodes along the path from the source to the destination as requested by the QoS session. QoS resource reservation mechanisms can be broadly classified into two categories: hard-state and soft-state reservation mechanisms.

In hard-state resource reservation schemes, resources are reserved at all intermediate nodes along the path from the source to the destination throughout the duration of the QoS session. If such a path is broken due to network dynamics, these reserved resources have to be released explicitly released by a deallocation mechanism. Such a mechanism not only introduces additional control overhead, but also may fail to release resources completely in case a node previously belonging to the session becomes unreachable. Due to these problems, soft-state resource reservation mechanisms, which maintain reservations only for small time intervals, are used. These reservations get refreshed if packets belonging to the same flow are received before the time-out period.

The soft-state reservation time-out period can be equal to packet interarrival time or a multiple of the packet interarrival time. If no data packets are received for the specified time interval, the resources are deallocated in a decentralized manner without incurring any additional control overhead. Thus, no explicit teardown is required for a flow. The hard-state schemes reserve resources explicitly; hence, at high network loads, the call-blocking ratio will be high, whereas soft-state schemes provide high call acceptance in a gracefully degraded fashion.

6.2.2 Stateful versus Stateless Approach

In the stateful approach, each node maintains either global state information or only local state information; in the case of the stateless approach, no such information is maintained at the nodes. State information includes both the topology information and the flow-specific information. The source node can use a centralized routing algorithm to route packets to the destination if the global state information is available. The performance of the routing protocol depends on the accuracy of the global state information maintained at the nodes. Significant control overhead is incurred in gathering and maintaining global state information.

On the other hand, if mobile nodes maintain only local state information (which is more accurate), distributed routing algorithms can be used. Even though control overhead incurred in maintaining local/ state information is low, care must be taken to obtain loop-free routes. In the case of the stateless approach, neither flow-specific nor link-specific state information is maintained at the nodes. Though the stateless approach solves the scalability problem permanently and reduces the burden (storage and computation) on nodes, providing QoS guarantees becomes extremely difficult.

6.2.3 Hard QoS versus Soft QoS Approach

The QoS provisioning approaches can be broadly classified into two categories: hard QoS and soft QoS approaches. If QoS requirements of a connection are guaranteed to be met for the whole duration of the session, the QoS approach is termed as a hard QoS approach. If the

QoS requirements are not guaranteed for the entire session, the QoS approach is termed as a soft QoS approach.

Keeping network dynamics of MANETs in mind, it is very difficult to provide hard QoS guarantees to user applications. Thus, QoS guarantees can only be given within certain statistical bounds. Almost all QoS approaches available in the literature provide only soft QoS guarantees.

6.3 Classification of QoS Solutions

Based on the interaction between the routing protocol and the MAC (media access control) protocol, QoS approaches can be classified into two categories: independent and dependent QoS approaches. In the independent QoS approach, the network layer is not dependent on the MAC layer for QoS provisioning. The dependent QoS approach requires the MAC layer to assist the routing protocol for QoS provisioning. Finally, based on the routing information, update mechanisms are employed.

6.3.1 MAC Layer Solutions

The MAC protocol determines which node should transmit next on the broadcast channel when several nodes are competing for transmission on that channel. Some of the MAC protocols that provide QoS support for applications in MANETs are described below.

6.3.1.1 Cluster TDMA Gerla and Tsai proposed cluster TDMA [1] for supporting real-time traffic in ad hoc wireless networks (AWNs). In bandwidth-constrained MANETs, the limited resources available need to be managed efficiently. To achieve this goal, a dynamic clustering scheme is used in cluster TDMA (time division multiple access). The available nodes in the network are split into different groups. Each group has a cluster head (elected by members of that group), which acts as a regional broadcast node and as a local coordinator to enhance the channel throughput. Every node within a cluster is one hop away from the cluster head. Formation of clusters and selection of cluster heads are done in a distributed manner. Clustering algorithms split the nodes into clusters such that they are interconnected and cover all the nodes. Three such algorithms used are the lowest ID algorithm, highest degree (degree refers to number of neighbors within transmission range of a node) algorithm, and least cluster change (LCC) algorithm.

In the lowest ID algorithm, a node becomes a cluster head if it has the lowest ID among all its neighbors. In the highest degree algorithm, a node with a degree greater than the degrees of all its neighbors becomes the cluster head. In the LCC algorithm, the cluster head change occurs only if a change in network causes two cluster heads to come into one cluster or one of the nodes moves out of the range of all the cluster heads. In each cluster, the corresponding cluster head maintains a power gain 2 matrix. It contains the power gain lists of all the nodes that belong to a particular cluster. It is useful for controlling the transmission power and the code division within a cluster.

The TDMA scheme is used within a cluster for controlling access to the channel. Further, it is possible for multiple sessions to share a given TDMA slot via CDMA (code division multiple access). Across clusters, either spatial reuse of the time slots or different spreading codes can be used to reduce the effect of intercluster interference. A synchronous time division frame is defined to support TDMA access within a cluster and to exchange control information.

Each synchronous time division frame is divided into slots. Slots and frames are synchronized throughout the network. A frame is split into a control phase and a data phase.

The data phase supports both real-time and best-effort traffic. Based on the bandwidth requirement of the real-time session, a virtual circuit (VC) is set up by allocating sufficient numbers of slots in the data phase. The remaining data slots (i.e., free slots) can be used by the best-effort traffic using the slotted-ALOHA scheme. For each node, a predefined slot is assigned in the control phase to broadcast its control information. The control information is transmitted over a common code throughout the network.

At the end of the control phase, each node will have learned, from the information broadcast by the cluster head, the slot reservation status of the data phase and the power gain lists of all its neighbors. This information helps a node to schedule free slots, verify the failure of reserved slots, and drop expired real-time packets. A fast reservation scheme is used in which a reservation is made when the first packet is transmitted, and the same slots in the subsequent frames can be used for the same connection. If the reserved slots remain idle for a certain time-out period, then they are released.

6.3.2 Network Layer Solutions

The bandwidth reservation and real-time traffic support capability of MAC protocols can ensure reservation at the link level only; hence, the network layer support for ensuring end-to-end resource negotiation, reservation, and reconfiguration is very essential.

To assist QoS routing, the topology information can be maintained at the nodes of AWNs. The topology information needs to be refreshed frequently by sending link-state update messages, which consume precious network resources such as bandwidth and battery power. Otherwise, the dynamically varying network topology may cause the topology information to become imprecise. This trade-off affects the performance of the QoS routing protocol. As path breaks occur frequently in AWNs compared to wired networks where a link goes down very rarely, the path satisfying the QoS requirements needs to be recomputed every time the current path gets broken. The QoS routing protocol should respond quickly in the case of path breaks and recompute the broken path or bypass the broken link without degrading the level of QoS.

6.4 QoS-Enabled Ad Hoc On-Demand Distance Vector Routing Protocol

Perkins, Royer, and Das [2] have extended the basic ad hoc on-demand distance vector (AODV) routing protocol to provide QoS support in AWNs. To provide QoS, packet formats have been modified in order to specify the service requirements that must be met by the nodes forwarding a route request (RREQ) or a route reply (RREP).

6.4.1 QoS Extensions to AODV Protocol

Each routing table entry corresponds to a different destination node. The following fields are appended to each routing table entry: maximum delay, minimum available bandwidth, list of sources requesting delay guarantees, and list of sources requesting bandwidth guarantees.

6.4.1.1 Maximum Delay Extension Field The maximum delay extension field is interpreted differently for RREQ and RREP messages. In an RREQ message, it indicates the maximum time (in seconds) allowed for a transmission from the current node to the destination node. In an RREP message, it indicates the current estimate of cumulative delay from the current intermediate node forwarding the RREP to the destination.

Using this field, the source node finds a path (if it exists) to the destination node satisfying the maximum delay constraint. Before forwarding the RREQ, an intermediate node compares its *node traversal time* (i.e., the time it takes for a node to process a packet) with the (remaining) delay indicated in the maximum delay extension field. If the delay is less than node traversal time, the node discards the RREQ packet. Otherwise, the node subtracts node traversal time from the delay value in the extension and processes the RREQ as specified in the AODV protocol.

The destination node returns an RREP with the maximum delay extension field set to zero. Each intermediate node forwarding the RREP adds its own node traversal time to the delay field and forwards the RREP toward the source. Before forwarding the RREP packet, the intermediate node records this delay value in the routing table entry for the corresponding destination node.

6.4.1.2 Minimum Bandwidth Extension Field Similarly, a minimum bandwidth extension field is also proposed to find a path (if it exists) to the destination node satisfying the minimum bandwidth constraint. A QOSLOST message is generated when an intermediate node experiences an increase in node traversal time or a decrease in the link capacity. The QOSLOST message is forwarded to all sources potentially affected by the change in the QoS parameter.

6.4.2 Advantages and Disadvantages

The advantage of the QoS AODV protocol is the simplicity of extension of the AODV protocol that can potentially enable QoS provisioning. But, as no resources are reserved along the path from the source to the destination, this protocol is not suitable for applications that require hard QoS guarantees. Further, node traversal time is only the processing time for the packet; the major part of the delay at a node is contributed by packet queuing and contention at the MAC layer. Hence, a packet may experience much more delay than this when the traffic load is high in the network.

6.5 QoS Frameworks for Ad Hoc Wireless Networks

A framework for QoS is a complete system that attempts to provide required/promised services to each user or application. All components within this system cooperate together in providing the required services. The key component of any QoS framework is the QoS model that defines the way user requirements are met. The key design issue here is whether to serve users on a per-session basis or on a per-class basis. Each class represents an aggregation of users based on certain criteria.

The other key components of the framework are QoS routing, which is used to find all or some of the feasible paths in the network that can satisfy user requirements; QoS signaling for resource reservation; QoS medium access control; call admission control; and packet scheduling schemes. The QoS modules should react promptly to changes in the network state (topology changes) and flow state (change in the end-to-end view of the service delivered).

The functionality of each component and its role in providing QoS in MANETs are described below:

- *Routing protocol.* The routing protocol is used to find a path from the source to the destination and to forward the data packet to the next intermediate relay node. The routing protocol needs to work efficiently with other components of the QoS framework in order to provide end-to-end QoS guarantees. These mechanisms should consume minimal resources in operation and react rapidly to changes in the network state and flow state.
- *QoS resource reservation signaling.* Once a QoS path is found, the resource reservation signaling protocol reserves the required

resources along that path. For example, for applications that require certain minimum bandwidth guarantees, a signaling protocol communicates with the MAC subsystem to find and reserve the required bandwidth. On completion/termination of a session, the previously reserved resources are released.

- *Admission control.* Even though a QoS feasible path may be available, the system needs to decide whether to serve the connection or not. If the call is to be served, the resources are reserved by the signal protocol; otherwise, the application is notified of the rejection. When a new call is accepted, it should not jeopardize the QoS guarantees given to the already admitted calls. A QoS framework is evaluated based on the number of QoS sessions it serves and it is represented by the ACAR metric. Admission control ensures that there is no perceivable degradation in the QoS being offered to the QoS sessions admitted already.

- *Packet scheduling.* When multiple QoS connections are active at the same time through a link, the decision on which QoS flow is to be served next is made by the scheduling scheme. For example, when multiple delay-constrained sessions are passing through a node, this module decides on when to schedule the transmission of packets when packets belonging to more than one session are pending in the transmission queue of the node. The performance of a scheduling scheme is reflected by the percentage of packets that meet their deadlines.

6.5.1 QoS Models

A QoS model defines the nature of service differentiation. In wired network QoS frameworks, several service models have been proposed. Two of these models are the integrated services (IntServ) model [3] and the differentiated services (DiffServ) model. The IntServ model provides QoS on a per-flow basis. The volume of information maintained at an IntServ-enabled router is proportional to the number of flows. Hence, the IntServ model is not scalable for the Internet, but it can be applied to small MANETs. But, per-flow information is difficult to maintain precisely at a node in an ad hoc wireless network. The DiffServ model was proposed in order to solve the scalability problem faced by the IntServ model. In this model, flows are aggregated into limited numbers of service classes. Each flow belongs to one of the DiffServ classes of service.

These two service models cannot be directly applied to MANETs because of unique characteristics such as continuously varying network topology, limited resource availability, and error-prone shared radio channel. Any service model proposed should first decide upon what types of services are feasible in such networks. A hybrid service model for MANETs called FQMM is described below. This model is based on these two QoS models.

6.5.1.1 Flexible QoS Model for Mobile Ad Hoc Networks The flexible QoS model for mobile ad hoc networks (FQMM) takes advantage of the per-flow granularity of IntServ and aggregation of services into classes in DiffServ. A source node, which is the originator of the traffic, is responsible for traffic shaping. Traffic shaping is the process of delaying packets belonging to a flow so that packets conform to a certain defined traffic profile. The traffic profile contains a description of the temporal properties of a flow such as its mean rate (i.e., rate at which data can be sent per unit time on average) and burst size (which specifies in bits per burst how much traffic can be sent within a given unit of time without creating scheduling concerns).

The FQMM model provides per-flow QoS guarantees for the high-priority flows while lower priority flows are aggregated into a set of service classes, as illustrated in Figure 6.1. This hybrid QoS model is based on the assumption that the percentage of flows requiring per-flow QoS guarantees is much less than that of low-priority flows, which can be aggregated into a set of QoS classes. Based on the current traffic load in the network, service level of a flow may change dynamically from per flow to per class and vice versa.

6.5.1.1.1 Advantages and Disadvantages This model addresses the scalability problem by classifying the low-priority traffic into service classes and it provides the ideal per-flow QoS guarantees. This protocol addresses the basic problem faced by QoS frameworks and proposes a generic solution for MANETs that can be a base for a better QoS model. But issues such as decision upon traffic classification, allotment

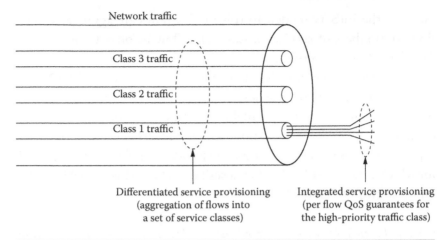

Figure 6.1 FQMM model.

of per-flow or aggregated service for the given flow, amount of traffic belonging to per-flow service, the mechanisms used by the intermediate nodes to get information regarding the flow, and scheduling or forwarding of the traffic by the intermediate nodes are as yet unresolved.

6.6 INSIGNIA

The INSIGNIA QoS framework was developed for providing adaptive services in MANETs. Adaptive services support applications that require only a minimum quantitative QoS guarantee (such as minimum bandwidth), called base QoS. The service level can be extended later to enhanced QoS when sufficient resources become available. Here, user sessions adapt to the available level of service without explicit signaling between the source–destination pairs.

This framework can scale down, drop, or scale up user sessions adaptively based on network dynamics and user-supplied adaptation policies. A key component of this framework is the INSIGNIA in-band signaling system, which supports fast reservation, restoration, and adaptation schemes to deliver the adaptive services. The signaling system is lightweight and responds rapidly to changes in the network topology and end-to-end QoS conditions. The INSIGNIA framework is depicted in Figure 6.2.

The routing module is independent of other components and hence any existing routing protocol can be used. INSIGNIA assumes that

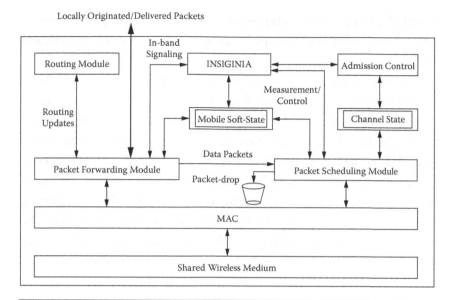

Figure 6.2 INSIGINIA QoS framework.

the routing protocol provides new routes in case of topology changes. The in-band signaling module is used to establish, adapt, restore, and tear down adaptive services between source–destination pairs. It is not dependent on any specific link layer protocol. In in-band signaling systems the control information is carried along with data packets and hence no explicit control channel is required.

In the INSIGNIA framework, each data packet contains an optional QoS field (INSIGNIA option) to carry the control information. The signaling information is encoded into this optional QoS field. The in-band signaling system can operate at speeds close to those of packet transmissions and is therefore better suited for highly dynamic mobile network environments. The admission control module uses the soft-state approach to allocate bandwidth to flows based on the maximum/minimum bandwidth requested.

The packet forwarding module classifies the incoming packets and delivers them to the appropriate module. If the packet has an INSIGNIA option, it is delivered to the INSIGNIA signaling module. Packets that are to be routed to other nodes are handled by the packet-scheduling module. The packets to be transmitted by a node are scheduled by the scheduler based on the forwarding policy. INSIGNIA uses a weighted round-robin service discipline. The

INSIGNIA framework is transparent to any underlying MAC protocol and uses a soft-state resource management mechanism for efficient
utilization of resources.

When an intermediate node receives a data packet with the RES
(reservation) flag set for a QoS flow and no reservation has been done
until now, the admission control module allocates the resources based
on availability. If the reservation has been done already, it is recon-
firmed. If no data packets are received for a specified time-out period,
the resources are deallocated in a distributed manner without incurring any control overhead. In setting the value for the time-out period,
care should be taken to avoid false restoration (which occurs when the
time interval is smaller than interarrival time of packets) and resource
lockup (which occurs when the time interval is much greater than
interarrival time of packets).

6.6.1 Operation of INSIGNIA Framework

The INSIGNIA framework supports adaptive applications, which can
be applications requiring best-effort service or applications with base
QoS requirements or those with enhanced QoS requirements. Due to
the adaptation of the protocol to the dynamic behavior of AWNs, the
service level of an application can be degraded in a distributed manner
if enough resources are not available.

The INSIGNIA option field contains the following information:
service mode, payload type, bandwidth indicator, and bandwidth
request. These indicate the dynamic behavior of the flow and the
requirements of the application. The intermediate nodes take decisions regarding the flow state in a distributed manner based on the
INSIGNIA option field. The service mode can be either best-effort
(BE) or service requiring reservation (RES) of resources. The payload type indicates the QoS requirements of the application. It can be
either base QoS for an application that requires minimum bandwidth
or enhanced QoS for an application requiring a certain maximum
bandwidth but that can operate with a certain minimum bandwidth
below which they are useless. Examples of applications that require
enhanced service mode are video applications that can tolerate packet
loss and delay jitter to a certain extent.

Table 6.1 How Service Mode, Payload Type, and Bandwidth Indicator Flags Reflect the Current Status of Flows

SERVICE MODE	PAYLOAD TYPE	BW INDICATOR	DEGRADING	UPGRADING
BE	–	–	–	–
RES	Base QoS	MIN	Base QoS→Be	BE→Base QoS
RES	Enhanced	MAX	EQoS→BE	BE→EQoS
	QoS (EQoS)		EQoS→BQoS	BQoS→EQoS

The bandwidth indicator flag has a value of MAX or MIN, which represents the bandwidth available for the flow. Table 6.1 shows how service mode, payload type, and bandwidth indicator flags reflect the current status of flows. It can be seen from the table that the best-effort (BE) packets are routed as normal data packets. If QoS is required by an application, it can opt for base QoS, in which a certain minimum bandwidth is guaranteed. For that application, the bandwidth indicator flag is set to MIN. For enhanced QoS, the source sets the bandwidth indicator flag to MAX, but it can be downgraded at the intermediate nodes to MIN; the service mode flag is changed to BE from RES if sufficient bandwidth is not available. The downgraded service can be restored to RES if sufficient bandwidth becomes available. For enhanced QoS, the service can be downgraded to either BE service or RES service with base QoS. The downgraded enhanced QoS can be upgraded later, if all the intermediate nodes have the required (MAX) bandwidth.

Destination nodes actively monitor ongoing flows, inspecting the bandwidth indicator field of incoming packets and measuring the delivered QoS (for example, packet loss, delay, and throughput). Destination nodes send QoS reports (which contain information regarding the status of the ongoing flows) to source nodes.

Route maintenance. Due to host mobility, an ongoing session may have to be rerouted in case of a path break. The flow restoration process has to reestablish the reservation as quickly and efficiently as possible. During restoration, INSIGNIA does not preempt resources from the existing flows for admitting the rerouted flows. INSIGNIA supports three types of flow restoration: immediate restoration, which occurs when a rerouted flow immediately recovers its original reservation; degraded restoration, which occurs when a rerouted flow is degraded for a period (T) before it recovers its original reservation; and permanent restoration, which occurs when the rerouted flow never recovers its original reservation.

6.6.2 Advantages and Disadvantages

The INSIGNIA framework provides an integrated approach to QoS provisioning by combining in-band signaling, call admission control, and packet scheduling together. The soft-state reservation scheme used in this framework ensures that resources are quickly released at the time of path reconfiguration. But, this framework supports only adaptive applications—for example, multimedia applications. Since this framework is transparent to any MAC protocol, the fairness and reservation scheme of the MAC protocol have a significant influence in providing QoS guarantees.

Also, as this framework assumes that the routing protocol provides new routes in the case of topology changes, the route maintenance mechanism of the routing protocol employed significantly affects the delivery of real-time traffic. If enough resources are not available because of the changing network topology, the enhanced QoS application may be downgraded to base QoS or even to best-effort service. As this framework uses in-band signaling, resources are not reserved before the actual data transmission begins.

Hence, INSIGNIA is not suitable for real-time applications that have stringent QoS requirements.

6.7 INORA

INORA is a QoS framework for MANETs that makes use of the INSIGNIA in-band signaling mechanism and the TORA (temporally ordered routing algorithm) routing protocol. The QoS resource reservation signaling mechanism interacts with the routing protocol to deliver QoS guarantees.

The TORA routing protocol provides multiple routes between a given source–destination pair. The INSIGNIA signaling mechanism provides feedback to the TORA routing protocol regarding the route chosen and asks for alternate routes if the route provided does not satisfy the QoS requirements. For resource reservation, a soft-state reservation mechanism is employed. INORA can be classified into two schemes: the coarse feedback scheme and the class-based fine feedback scheme.

6.7.1 Coarse Feedback Scheme

In this scheme, if a node fails to admit a QoS flow either due to lack of minimum required bandwidth (BWmin) or because of congestion at the node, it sends an out-of-band admission control failure (ACF) message to its upstream node. After receiving the ACF message, the upstream node reroutes the flow through another downstream node provided by the TORA routing protocol. If none of its neighbors are able to admit the flow, it in turn sends an ACF message to its upstream node.

While INORA is trying to find a feasible path by searching the directed acyclic graph (DAG) following admission control failure at an intermediate node, the packets are transmitted as best-effort packets from the source to its destination. In this scheme, different flows between the same source–destination pair can take different routes.

6.7.2 Class-Based Fine Feedback Scheme

In this scheme, the interval between BWmin and BWmax of a QoS flow is divided into N classes, where BWmin and BWmax are the minimum and maximum bandwidths required by the QoS flow. Consider a QoS flow being initiated by the source node S to destination node D. Let the flow be admitted with class m (m < N).

1. Let the DAG created by the TORA protocol be as shown in Figure 6.3. Let S→A→B→D be the path chosen by the TORA routing protocol.
2. INSIGNIA tries to establish soft-state reservations for the QoS flow along the path. Assume that node A has admitted the flow with class m successfully and node B has admitted the flow with bandwidth of class l (l < m) only.

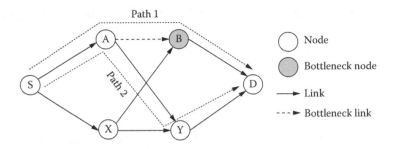

Figure 6.3 INORA fine feedback scheme. Node A has admitted the flow with class m, but node B is able to give it class l (l < m).

3. Node B sends an admission report message (AR(l)) to upstream node A, indicating its ability to give only class l bandwidth to the flow.
4. Node A splits the flow in the ratio of l to ml and forwards the flow to node B and node Y, in that ratio.
5. If node Y is able to give class (ml) as requested by node A, then the flow of class m is split into two flows: one flow with bandwidth of class l along the path S→A→B→D and the other with bandwidth of class (ml) along path S→A→Y→D.
6. If node Y gives only class n (n < ml), it sends an AR(n) message to the upstream node A.
7. Node A, realizing that its downstream neighbors are unable to give class m service, informs its ability to provide service class of (l + n) by sending an AR(l + n) to node S.
8. Node S tries to find another downstream neighbor that might be able to accommodate the flow with class (m(l + n)).
9. If no such neighbor is available, node S rejects the flow.

6.7.3 Advantages

INORA is better than INSIGNIA in that it can search multiple paths with fewer QoS guarantees. It uses the INSIGNIA in-band signaling mechanism. Since no resources are reserved before the actual data transmission begins and since data packets have to be transmitted as best-effort packets in case of admission control failure at the intermediate nodes, this model may not be suitable for applications that require hard-service guarantees.

6.8 Summary

The increased interest in MANETs in recent years has led to intensive research efforts that aim to provide QoS support over such infrastructure-less networks with unpredictable behavior. Generally, the QoS of any particular network can be defined as its ability to deliver a guaranteed level of service to its users and/or applications. These service requirements often include performance metrics such as throughput, delay, jitter (delay variance), bandwidth, reliability, etc., and different applications may have varying service requirements. The performance metrics can be computed in three different ways: (1) concave (e.g., minimum bandwidth along each link), (2) additive (e.g., total delay along a path), and (3) multiplicative (e.g., packet delivery ratio along the entire route).

In MANETs, the provision of QoS guarantees is much more challenging than in wire-line networks, mainly due to node mobility, multihop communications, contention for channel access, and a lack of central coordination. QoS guarantees are required by most multimedia and other time- or error-sensitive applications. The difficulties in the provision of such guarantees have limited the usefulness of MANETs. However, in the last decade, much research attention has focused on providing QoS assurances in MANET protocols. The QoS routing protocol is an integral part of any QoS solution since its function is to ascertain which nodes, if any, are able to serve applications' requirements. Consequently, it also plays a crucial role in data session admission control.

Problems

6.1 Quality of service is needed at all layers. Justify.

6.2 Describe the challenges and issues involved in providing QoS.

6.3 Give the classifications of QoS solutions.

6.4 Explain different types of QoS models with suitable illustrations.

6.5 Explain the FQMM model.

6.5 Discuss the INSIGNIA framework QoS model.

6.7 Describe the INORA framework with a suitable example.

References

1. IEEE 802.11 TGe. 2001. HCF ad hoc group recommendation—Normative text to EDCF access category. TR-02/241r0.
2. Perkins, C. E., E. M. Royer, and S. R. Das. 2000. Quality of service for ad hoc on-demand distance vector routing (work in progress). IETF Internet draft (draft-ietf-manet-aodvqos-00.txt).

Bibliography

Bheemarjuna Reddy, T., I. Karthigeyan, B. S. Manoj, and C. Siva Ram Murthy. 2006. Quality of service provisioning in ad hoc wireless networks: A survey of issues and solutions. *Ad Hoc Networks* 4:83–124.

Chen, S., and K. Nahrstedt. 1999. Distributed quality-of-service routing in ad hoc networks. *IEEE Journal on Selected Areas in Communications* 17 (8):1488–1504.

Chen, Y., Y. Tseng, J. Sheu, and P. Kuo. 2002. On-demand, link state, multi-path QoS routing in a wireless mobile ad-hoc network. *Proceedings of European Wireless 2002*, pp. 135–141.

De, S., S. K. Das, H. Wu, and C. Qiao. 2002. Trigger-based distributed QoS routing in mobile ad hoc networks. *ACM SIGMOBILE Mobile Computing and Communications Review* 6 (3):22–35.

IEEE 802.11 TGe. 2001. EDCF proposed draft text. TR-01/131r1, March 2001.

———. 2001. Hybrid coordination function (HCF)—Proposed updates to normative text of D0.1. TR-01/110r1, March 2001.

———. 2003. Proposed normative text for AIFS—Revisited. TR-01/270r0, February 2003.

Lin, C. R. 2001. On-demand QoS routing in multihop mobile networks. *Proceedings of IEEE INFOCOM 2001* 3:1735–1744.

Lin, C. R., and M. Gerla. 1999. Real-time support in multihop wireless networks. *Wireless Networks* 5 (2):125–135.

Lin, C. R., and J. Liu. 1999. QoS routing in ad hoc wireless networks. *IEEE Journal on Selected Areas in Communications* 17 (8):1426–1438.

Mangold, S., S. Choi, P. May, O. Klein, G. Hiertz, and L. Stibor. 2002. IEEE 802.11e wireless LAN for quality of service. *Proceedings of the European Wireless 2002* 1:32–39.

Perkins, C. E., and P. Bhagwat. 1994. Highly dynamic destination sequenced distance-vector routing (DSDV) for mobile computers. *Proceedings of ACM SIGCOMM 1994* 24 (4):234–244.

Perkins, C. E., and E. M. Royer. 1999. Ad hoc on-demand distance vector routing. *Proceedings of IEEE Workshop on Mobile Computing Systems and Applications*, February, pp. 90–100.

Shah, S. H., and K. Nahrstedt. 2002. Predictive location-based QoS routing in mobile ad hoc networks. *Proceedings of IEEE ICC 2002* 2:1022–1027.

Sheu, S., and T. Sheu. 2001. DBASE: A distributed bandwidth allocation/sharing/extension protocol for multimedia over IEEE 802.11 ad hoc wireless LAN. *Proceedings of IEEE INFOCOM 2001* 3:1558–1567.

Vidhyashankar, V., B. S. Manoj, and C. Siva Ram Murthy. 2003. Slot allocation schemes for delay sensitive traffic support asynchronous wireless mesh networks. *Proceedings of HiPC 2003,* December 2003.

7

ENERGY MANAGEMENT SYSTEMS

7.1 Introduction

A mobile ad hoc network (MANET) is a collection of digital data terminals that can communicate with one another without any fixed networking infrastructure. Since the nodes in a MANET are mobile, the routing and power management become critical issues. Wireless communication has the advantage of allowing untethered communication, which implies reliance on portable power sources such as batteries. However, due to the slow advancement in battery technology, battery power continues to be a constrained resource, so power management in wireless networks remains an important issue.

Though many proactive and reactive routing protocols exist for MANETs, the reactive dynamic source routing (DSR) protocol is considered an efficient protocol. But, when the network size is increased, it is observed that in DSR overhead and power consumption of the nodes in the network increase, which in turn drastically reduces the efficiency of the protocol.

7.1.1 Why Energy Management Is Needed in Ad Hoc Networks

The energy management in ad hoc networks is a very important aspect of the overall management of ad hoc networks. The mobile wireless sensor nodes in the field need to conserve energy and use it optimally in order to play the assigned role in an ad hoc network for a longer period of time

Battery power is an important resource in ad hoc networks. It has been observed that in these networks, energy consumption does not reflect the communication activities in the network. Many existing energy conservation protocols based on electing a routing backbone for global connectivity are oblivious to traffic characteristics.

Various techniques, both in hardware and software, have been proposed to reduce energy consumption for mobile computing devices in wireless LANs [4,5]. In contrast, power management in ad hoc networks is a more difficult problem for two reasons. First, in ad hoc networks, a node can be both a data source/sink and a router that forwards data for other nodes and participates in high-level routing and control protocols. Additionally, the roles of a particular node may change over time. Second, there is no centralized entity such as an access point to control and maintain the power management mode of each node in the network, buffer data, and wake up sleeping nodes. Therefore, power management in ad hoc networks must be done in a distributed and cooperative fashion. A major challenge to the design of a power management framework for ad hoc networks is that energy conservation usually comes at the cost of degraded performance, such as lower throughput or longer delay. A naive solution that only considers power savings at individual nodes may turn out to be detrimental to the operation of the whole network.

7.1.2 Classification of Energy Management Schemes

The nodes in an ad hoc wireless network (AWN) are constrained by limited battery power for their operation. Hence, energy management is an important issue in such networks. The use of multi-hop radio relaying requires a sufficient number of relaying nodes to maintain network connectivity. Hence, battery power is a precious resource that must be used efficiently in order to avoid early termination of any node. Energy awareness thus needs to be adopted by the protocols at all the layers in the protocol stack, and has to be considered one of the important design objectives for all the protocols in AWNs. Most energy management solutions for AWNs follow similar methodologies to increase the network lifetime, and we classify them as follows:

- Battery management schemes
- Transmission power management schemes
- System power management schemes

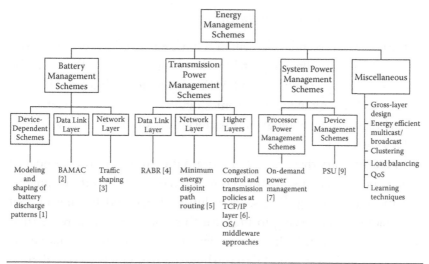

Figure 7.1   Taxonomy of energy management schemes.

Figure 7.1 shows a schematic diagram of these classifications and lists an example under each of them. Maximizing the life of an AWN requires an understanding of the capabilities and limitations of energy sources of the nodes. Greater battery capacity leads to a longer lifetime of the nodes. Battery management is concerned with problems that lie in the selection of battery technologies, finding the optimal battery capacity, and scheduling the batteries to increase capacity. Transmission power management techniques attempt to find an optimum transmission range for the nodes in the AWN.

System power management, on the other hand, deals mainly with minimizing the power required by hardware peripherals of a node and incorporating low-power strategies into the protocols used in various layers of the protocol stack. It can be further divided into device and processor power management schemes. Although the energy management schemes for AWNs cannot be strictly classified under the different layers of the open systems interconnection (OSI) protocol stack, as they reside in more than one layer, the classification provided in this chapter is based on the highest layer in the protocol stack used by each of these protocols. A few other techniques, which do not fall into any of these categories, are classified as miscellaneous schemes.

7.1.3 Overview of Battery Technologies

Batteries are an essential element of today's electronics scene. Batteries are used in virtually all portable electronics devices, from mobile phones to laptop computers and MP3 players to flashlights. Without battery technology, many electronics devices would not be viable. As a result, battery technology and battery development are essential to today's electronics.

In recent years there has been a dramatic growth in the number of battery-powered items and this has resulted in many new developments in battery technology. The sheer volume of demand has meant that manufacturers are trying to improve their products to increase their share of the market. If they can achieve this, enormous returns can be made on their investment. With the huge demand for batteries, there is a wide variety of different battery and cell technologies available. These range from the established nonrechargeable technologies such as zinc-carbon and alkaline batteries to rechargeable batteries that have moved from nickel cadmium (NiCd) through nickel metal hydride (NiMH) cells to the newer lithium ion rechargeable batteries. With a huge need for batteries, there is a large amount of battery technology development under way, and new types of cells and batteries offering even higher levels of performance will no doubt become available.

Another area of battery technology that is becoming more important is the green or environmental aspects. Some of the old battery technologies contain chemicals which can be considered toxic. Now, new designs are seeking to use more environmentally friendly chemicals. Nickel cadmium cells are now considered environmentally unfriendly and are not as widely used as they were previously. Other batteries also contain harmful chemicals, and this is likely to have a significant impact on the direction of future developments.

There are different types of batteries:

- **Nickel cadmium (NiCd) batteries and cells** have been widely used in applications where electrical rechargeable power sources are needed. These NiCd cells have been used for many applications where electronic equipment such as laptop computers, electronic games, mobile phones, and many other items of electronics equipment have needed a form of rechargeable power source. In addition to this, nickel cadmium cells have also been widely used for flashlights and other small items of electronic equipment. NiCd cells are less widely used these days because of their use of cadmium, which has to be disposed of carefully when the battery life has been finished. These environmental concerns, along with the fact that there are more efficient cells available, have brought about a decline in the use of nickel cadmium cells.

- **Nickel metal hydride (NiMH) batteries and cells** have come into widespread use in recent years as a viable form of rechargeable battery. These cells offer almost identical characteristics to those provided by the older NiCd technology, but with the advantage that the NiMH cells do not have the same adverse environmental effects, and they are also able to provide a slightly higher level of energy density and therefore overall charge capacity. As a result, NiMH cells are now widely used, offering high levels of performance.

- **Lithium ion (Li-ion) batteries** are now being widely used for applications such as powering laptop computers, mobile phones, cameras, and many more devices. The high-energy density that Li-ion batteries provide enables the electronic devices they power to be recharged less frequently. Also, Li-ion batteries are comparatively light when compared to other forms of rechargeable cells and batteries. In view of their convenience, Li-ion batteries are widely used and there are a number of different manufacturers for these batteries. Accordingly, costs have fallen from their original high levels, although Li-ion batteries are still expensive.

7.1.4 Principles of Battery Discharge

The purpose of a battery is to store and release energy at the desired time and in a controlled manner. This section examines discharges under different C-rates (discharge rates) and evaluates the depth to which a battery can safely be depleted.

7.1.4.1 Depth of Discharge   The end-of-discharge voltage for lead acid is 1.75 V/cell, a nickel-based system is 1.00 V/cell, and most Li-ion systems are 3.00 V/cell. At this level, roughly 95% of the energy is

Table 7.1   Recommended End-of-Discharge Voltage under Normal and Heavy Loads

END OF DISCHARGE	LI-MANGANESE	LI-PHOSPHATE	LEAD ACID	NICD/NIMH
Normal load	3.00 V/cell	2.70 V/cell	1.75 V/cell	1.00 V/cell
Heavy load	2.70 V/cell	2.45 V/cell	1.40 V/cell	0.90 V/cell

spent and the voltage would drop rapidly if the discharge were to continue. To protect the battery from overdischarging, most devices prevent operation beyond the specified end-of-discharge voltage.

When removing the load after discharge, the voltage of a healthy battery gradually recovers and rises toward the nominal voltage. Differences in the metal concentration of the electrodes enable this voltage potential when the battery is empty. An aging battery with elevated self-discharge cannot recover the voltage because of the parasitic load.

A high load current lowers the battery voltage, so the end-of-discharge voltage threshold should be set lower accordingly. Internal cell resistance, wiring, protection circuits, and contacts all add up to overall internal resistance. The cutoff voltage should also be lowered when discharging at very cold temperatures; this compensates for the higher than normal internal resistance. Table 7.1 shows typical end-of-discharge voltages of various battery chemistries.

The lower end-of-discharge voltage on a high load compensates for the losses induced by the internal battery resistance.

Some battery analyzers apply a secondary discharge (recondition) that drains the battery voltage of a nickel-based battery to 0.5 V/cell and lower, a cutoff point that is below what manufacturers specify. These analyzers (Cadex) keep the discharge load low to stay within an allowable current while in subdischarge range. A cell breakdown with a weak cell is possible and reconditioning would cause further deterioration in performance rather than making the battery better. This phenomenon can be compared to the experience of a patient to whom strenuous exercise is harmful.

7.1.5 Impact of Discharge Characteristics on Battery Capacity

7.1.5.1 Temperature Characteristics   Cell performance can change dramatically with temperature. At the lower extreme, in batteries with

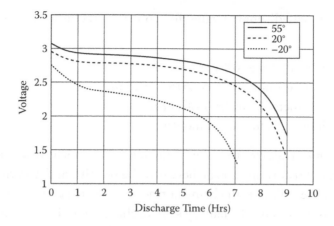

Figure 7.2 Performance of lithium ion batteries with temperature.

aqueous electrolytes, the electrolyte itself may freeze, setting a lower limit on the operating temperature. At low temperatures, lithium batteries suffer from lithium plating of the anode, causing a permanent reduction in capacity. At the upper extreme, the active chemicals may break down, destroying the battery. In between these limits, the cell performance generally improves with temperature.

Figure 7.2 shows how the performance of lithium ion batteries deteriorates as the operating temperature decreases. Probably more important is that, for both high and low temperatures, the further the operating temperature is from room temperature the more the cycle life is degraded.

7.1.5.2 Self-Discharge Characteristics The self-discharge rate is a measure of how quickly a cell will lose its energy while sitting on the shelf due to unwanted chemical actions within the cell. The rate depends on the cell chemistry and the temperature.

Cell chemistry. The following shows the typical shelf life for some primary cells:

- Zinc carbon (Leclanché): 2 to 3 years
- Alkaline: 5 years
- Lithium: 10 years or more

Typical self-discharge rates for common rechargeable cells are as follows:

- Lead acid: 4% to 6% per month
- Nickel cadmium: 15% to 20% per month
- Nickel metal hydride: 30% per month
- Lithium: 2% to 3% per month

Temperature effects. The rate of unwanted chemical reactions that cause internal current leakage between the positive and negative electrodes of the cell, like all chemical reactions, increases with temperature, thus increasing the battery self-discharge rate. Figure 7.3 shows typical self-discharge rates for a lithium ion battery.

Internal impedance. The internal impedance of a cell determines its current carrying capability. A low internal resistance allows high currents.

Battery equivalent circuit. The diagram in Figure 7.4 shows the equivalent circuit for an energy cell.

- **Rm** is the resistance of the metallic path through the cell, including the terminals, electrodes, and interconnections.
- **Ra** is the resistance of the electrochemical path, including the electrolyte and the separator.

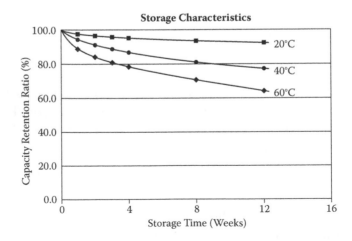

Figure 7.3   Self-discharge rates for a lithium ion battery.

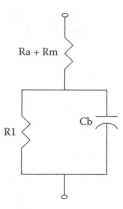

Figure 7.4   Equivalent circuit for an energy cell.

- **Cb** is the capacitance of the parallel plates that form the electrodes of the cell.
- **Ri** is the nonlinear contact resistance between the plate or electrode and the electrolyte.

Typical internal resistance is in the order of milliohms.

7.1.5.3 Effects of Internal Impedance   When current flows through the cell, there is an IR (current resistance) voltage drop across the internal resistance of the cell, which decreases the terminal voltage of the cell during discharge and increases the voltage needed to charge the cell, thus reducing its effective capacity as well as decreasing its charge/discharge efficiency. Higher discharge rates give rise to higher internal voltage drops; this explains the lower voltage discharge curves at high C-rates. (See Section 7.1.5.4.)

The internal impedance is affected by the physical characteristics of the electrolyte; the smaller the granular size of the electrolyte material is, the lower is the impedance. The grain size is controlled by the cell manufacturer in a milling process.

Spiral construction of the electrodes is often used to maximize the surface area and thus reduce internal impedance. This reduces heat generation and permits faster charge and discharge rates. The internal resistance of a galvanic cell is temperature dependent, decreasing as the temperature rises due to the increase in electron mobility. The following graph is a typical example.

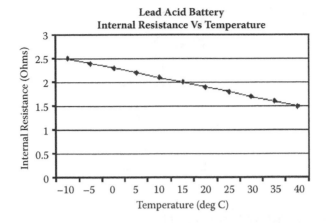

Thus, the cell may be very inefficient at low temperatures but the efficiency improves at higher temperatures due to the lower internal impedance as well as to the increased rate of the chemical reactions. However, the lower internal resistance unfortunately also causes the self-discharge rate to increase. Furthermore, cycle life deteriorates at high temperatures. Some form of heating and cooling may be required to maintain the cell within a restricted temperature range to achieve the optimum performance in high-power applications.

The internal resistance of most cell chemistries also tends to increase significantly toward the end of the discharge cycle as the active chemicals are converted to their discharged state and hence are effectively used up. This is principally responsible for the rapid drop in cell voltage at the end of the discharge cycle. In addition, the Joule heating effect of the I^2R losses in the internal resistance of the cell will cause the temperature of the cell to rise.

The voltage drop and the I^2R losses may not be significant for a 1000 mAh cell powering a mobile phone, but for a 100-cell, 200 Ah automotive battery they can be substantial. Typical internal resistance for a 1000 mA lithium mobile phone battery is around 100 to 200 mΩ and around 1 mΩ for a 200Ah lithium cell used in an automotive battery. Operating at the C-rate, the voltage drop per cell will be about 0.2 V in both cases (slightly less for the mobile phone). The I^2R loss in the mobile phone will be between 0.1 and 0.2 W. In the automotive battery, however, the voltage drop across the whole battery will be 20 V and I^2R power loss dissipated as heat within the battery will be 40

W per cell or 4 kW for the whole battery. This is in addition to the heat generated by the electrochemical reactions in the cells.

As a cell ages, the resistance of the electrolyte tends to increase. Aging also causes the surface of the electrodes to deteriorate; the contact resistance builds up and, at the same, the effective area of the plates decreases, reducing its capacitance. All of these effects increase the internal impedance of the cell, adversely affecting its ability to perform. Comparing the actual impedance of a cell with its impedance when it was new can be used to give a measure or representation of the age of a cell or its effective capacity. Such measurements are much more convenient than actually discharging the cell and can be taken without destroying the cell under test.

The internal resistance also influences the effective capacity of a cell. The higher the internal resistance is, the higher are the losses while charging and discharging, especially at higher currents. This means that, for high discharge rates, the available capacity of the cell is lower. Conversely, if it is discharged over a prolonged period, the amp hour capacity is higher. This is important because some manufacturers specify the capacity of their batteries at very low discharge rates, which makes them look a lot better than they really are.

7.1.5.4 Discharge Rates The discharge curves for a lithium ion cell in the following graph show that the effective capacity of the cell is reduced if the cell is discharged at very high rates (or, conversely, increased with low discharge rates). This is called the capacity offset and it is common to most cell chemistries.

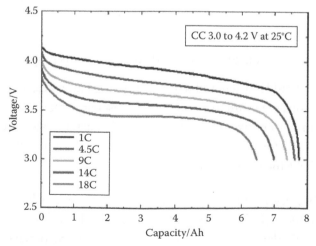

7.1.5.5 Battery Load Battery discharge performance depends on the load the battery has to supply. If the discharge takes place over a long period of several hours, as with some high-rate applications such as electric vehicles, the effective capacity of the battery can be as much as double the specified capacity at the C-rate. This can be most important when dimensioning an expensive battery for high-power use. The capacity of low-power consumer electronics batteries is normally specified for discharge at the C-rate, whereas the SAE (Society of Automotive Engineers) uses the discharge over a period of 20 hours (0.05 C) as the standard condition for measuring the amp hour capacity of automotive batteries. The following graph shows that the effective capacity of a deep discharge lead acid battery is almost doubled as the discharge rate is reduced from 1.0 to 0.05 C. For discharge times less than 1 hour (high C-rates) the effective capacity falls off dramatically. The effectiveness of charging is similarly influenced by the rate of charge.

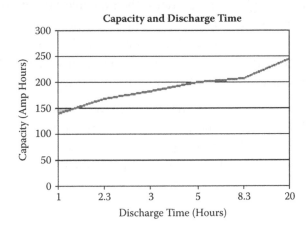

There are two conclusions to be drawn from this graph:

- Care should be exercised when comparing battery capacity specifications to ensure that comparable discharge rates are used.
- In an automotive application, if high current rates are used regularly for hard acceleration or for hill climbing, the range of the vehicle will be reduced.

7.1.5.6 Duty Cycle Duty cycles are different for each application. Electric vehicle (EV) and hybrid electric vehicle (HEV) applications impose particular, variable loads on the battery. Stationary batteries used in distributed grid energy storage applications may have very large system on chip (SOC) changes and many cycles per day. It is important to know how much energy is used per cycle and to design for the maximum energy throughput and power delivery, not the average.

7.1.6 Battery Modeling

Researchers around the world have developed a wide variety of models with varying degrees of complexity. They capture battery behavior for specific purposes, from battery design and performance estimation to circuit simulation. Electrochemical models [11–14], mainly used to optimize the physical design aspects of batteries, characterize the fundamental mechanisms of power generation and relate battery design parameters with macroscopic (e.g., battery voltage and current) and microscopic (e.g., concentration distribution) information. However, they are complex and time consuming because they involve a system of coupled time-variant spatial partial differential equations [13]—a solution that requires days of simulation time, complex numerical algorithms, and battery-specific information that is difficult to obtain because of the proprietary nature of the technology.

The field of battery modeling can be divided into the following two areas.

1. **Estimation of battery performance.** Given an already constructed battery, the problem is to estimate how that battery will perform under specific conditions of interest to the user of the battery. This problem is typically addressed by testing batteries under the specific conditions of interest and using a model to represent the test results. Approaches for representing test results range from simple statistical models to neural nets to complex, physics-based models. Basing the model on test data becomes problematical when testing becomes impractical (such as a 10- to 20-year life test). Real-time estimation of battery performance, an important problem in automotive applications, falls into this area.

2. **Battery design.** Here the problem is to estimate how the design of a battery impacts its performance. This is a difficult problem and can be only partly addressed because the

complexity of most battery systems defies characterization. Our inability to characterize the mechanisms involved in many battery chemistries limits the application of modeling to battery design. Instead, battery design relies heavily on the tried and true approach of build and test rather than on engineering principles. This build and test approach is practical because test cells are often inexpensive to build and key tests often can be carried out rapidly. In the short term, developing a battery by trial and error actually takes less time than determining how a battery works and using that mechanistic understanding for design. However, those aspects of battery operation that are understood well enough to model, such as temperature and current distribution, have undergone significant optimization. Such advances indicate that as our understanding of batteries increases and more aspects of battery operation become amenable to modeling, we may expect a dramatic acceleration in the pace of battery development.

The area of battery performance estimation receives much more attention than the area of battery design. Battery performance can be estimated by a wide range of workers, while the area of battery design is limited mostly to battery developers. Battery developers consider design information highly proprietary and are reluctant to divulge such information to model developers, who, for the most part, still tend to be academics. Recently available, third-party battery design software provides some standard designs that can be studied openly and thus promotes development of the science of battery design.

However, progress in the area of battery design has benefited most from the advent of lithium ion batteries. Lithium ion or rocking-chair batteries are the newest batteries and have the most well-understood battery chemistry. Both the positive and negative electrodes serve simply as hosts for lithium ions that transport through a binary electrolyte. This system can be readily modeled. For example, a recent paper [6] by a battery developer shows that a physics-based model can provide a remarkably accurate estimate of battery behavior. This understanding of lithium ion batteries has encouraged modelers to develop successful methodologies for design of charge/discharge performance and abuse tolerance (Figure 7.5).

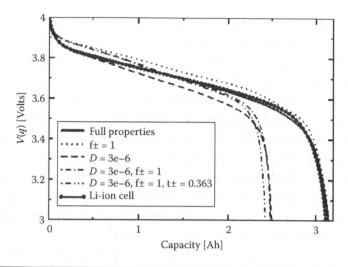

Figure 7.5 Comparison of simulated discharge curve to experimental discharge curve for a Li-ion cell.

7.1.7 Battery-Driven System Design

The activity of several components in a computing system is event driven. For example, the activity of display servers, communication interfaces, and user interface functions is triggered by external events, and it is often interleaved with long, idle periods. An intuitive way to reduce average power dissipated by the whole system consists of shutting down resources during periods of inactivity. In other words, one can adopt a dynamic power management (DPM) policy that dictates how and when various components should be shut down according to a system's workload.

Battery-operated portable appliances impose tight constraints on the power dissipation of their components. Such constraints are becoming tighter as complexity and performance requirements are pushed forward by user demand. Reducing power dissipation is a design objective also for stationary equipment, because excessive power dissipation implies increased cost and noise for complex cooling systems. Numerous computer-aided design techniques for low power have been proposed targeting digital, very large-scale integration (VLSI) circuits (i.e., chip-level designs). Almost every portable electronic appliance is far more complex than a single chip. Portable devices such as cellular telephones and laptop computers contain tens or even hundreds of components. To complicate the picture further, in most electronic products, digital components are responsible for only a fraction of the total power consumed. Analog, electromechanical, and optical components are often responsible for the largest contributions to the power budget.

One of the most successful techniques employed by designers at the system level is dynamic power management [8,9]. This technique reduces power dissipation by selectively turning off (or reducing the performance of) system components when they are idle (or partially unexploited). Building a complex system that supports dynamic power management is a difficult and error-prone process. Long trial-and-error iterations cannot be tolerated when fast time to market is the main factor deciding the success of a product.

To shorten the design cycle of complex power-managed systems, several hardware and software vendors are pursuing a long-term strategy to simplify the task of designing large and complex power-managed systems. The strategy is based on a standardization initiative known as the advanced configuration and power interface (ACPI). ACPI specifies an abstract and flexible interface between power-manageable hardware components (VLSI chips, disk drivers, display drivers, etc.) and the power manager (the system component that controls when and how to turn on and off functional resources). The ACPI interface specification simplifies the task of controlling the operating conditions of the system resources, but it does not provide insight on how and when to power manage them.

We call power-management policy ("policy" for brevity) a procedure that takes decisions upon the state of operation of system components and on the state of the system itself. The most aggressive policy (that we call eager policy) turns off every system component as soon as it becomes idle. Whenever the functionality of a component is required to carry out a system task, the component must be turned on and restored to its fully functional state. The transition between the inactive and the functional state requires time and power. As a result, the eager policy is often unacceptable because it degrades performance and may not decrease power dissipation.

For instance, consider a device that dissipates 2 W in a fully operational state and no power when set into an inactive state. The transition from an operational to an inactive state is almost instantaneous (hence, it does not consume sizable power). However, the opposite transition takes 2 s. During the transition, the power consumption is 4 W. This device is a highly simplified model of a hard-disk drive (a more detailed model will be introduced later in this chapter). Clearly, the eager policy does not produce any power savings if the device remains idle for less than 4 s. Moreover, even if the idle time is longer than 4 s, transitioning the device to inactive state degrades performance. If the eager policy is chosen, the user will experience a 2 s delay every time a request for the device is issued after an idle interval.

The choice of the policy that minimizes power under performance constraints (or maximizes performance under power constraint) is a constrained optimization problem that is of great relevance for low-power electronic systems. We call this problem policy optimization (PO). Several heuristic power-management policies have been investigated in the past, but no strong optimality result has been proven.

7.1.7.1 Stochastic Model Stochastic modeling-based approaches to DPM have been based on the framework of stationary discrete-time Markov chains, continuous-time Markov chains, or their variants. Irrespective of whether they are based on stationary discrete-time Markov chains, continuous-time Markov chains, or their variants, existing methodologies depend on modeling the input arrival process and the behavior of power-managed components by creating the stochastic matrices or generator matrices for these processes by hand, and then creating and solving optimization problems from those to optimize the average case.

One novelty of this work is the behavior of the input generator and power-managed component, as well as the power manager, in a high-level probabilistic language for expressing stochastic state machines. This allows automatic generation of the matrices; the rest of the required computation for designing strategies is then carried out in the model checking framework. In the power manager, managed components are modeled using stochastic petri nets. This allows automatic generation of the stochastic matrices and the formulation of the optimization problems. These exact optimization problems are meant to optimize the average energy usage while minimizing average delay. They are usually validated by simulation to check for the soundness of the modeling assumptions and effectiveness of the strategies in practice.

Since probabilistic model checking is inherently exhaustive in its search among all possible scenarios, more useful information can be obtained about the design space than using simulation. For example, optimal buffer sizes, average delays, probabilities of various corner case scenarios, etc, and probability-based comparisons between various delay-cost possibilities (obtainable by competing DPM strategies) can easily be predicted.

7.1.8 Smart Battery System

A smart battery system (SBS) is a specification for determining accurate battery capacity readings. It allows operating systems to perform power-management operations based on remaining estimated run times. The specifications to these smart battery systems were developed by Duracell and Intel in 1994 and later deployed by several battery and semiconductor manufacturing companies. The smart battery system specifications or standards are looked after by the SBS forum, whose main objective is to create an open standard that enables systems to be aware of the batteries that power them, improve battery efficiency, etc.

In a smart battery system, communication is carried over a system management bus (SMBus) two-wire communication bus. Through this communication, the system also controls the amount that a battery must be charged. The SBS specifies the following components:

1. **System management bus (SMBus).** The system management bus is a specific implementation of an I^2C bus that describes protocols for data, device addresses, and additional electrical requirements designed to transport commands and information physically between the smart battery, SMBus host, smart battery charger, and other smart devices.

2. **Battery data set.** The smart battery data (SBD) set is a method to monitor a rechargeable battery pack. A special integrated circuit (IC) in the battery pack monitors the battery and reports information to the SMBus. This information might include type, model number, manufacturer, characteristics,

discharge rate, predicted remaining capacity, almost discharged alarm so that the system can shut down gracefully, temperature, and voltage to provide safe fast-charging.

3. **Smart battery charger.** This battery charger periodically communicates with a smart battery and alters its charging characteristics in response to information provided by the smart battery.

4. **Smart battery system selector.** This device arbitrates between two or more smart batteries. It controls the power and SMBus paths between the SMBus host, a smart battery charger, and the smart batteries.

5. **Smart battery system manager.** The smart battery system manager is a specification that describes the requirements and the interface for a component or system of components that manages a number of smart batteries in a system.

Figure 7.6 shows a block diagram of a smart battery system. Smart battery A or/and B are available to power the system. The smart battery charger and SMBus act as an added functionality to the

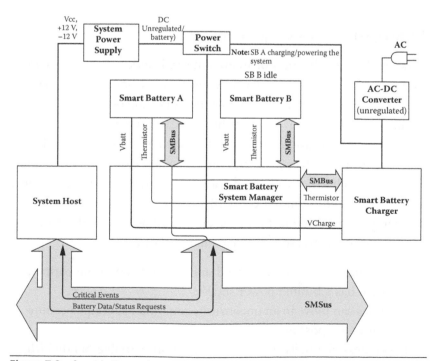

Figure 7.6 Smart battery system.

embedded controller. The power path configuration block is used by the SBSM to select which battery is used to power the system (i.e., to select smart battery A or smart battery B or a combination of both). The system's designer selects the algorithm used, which is contained in the SBSM. If alternating current (AC) is present, the SBSM may choose to charge either smart battery A or B or both. Again, the algorithm used is contained entirely within the SBSM. The safety signal combiner ensures that the smart battery charger's alternative safety signaling path is always maintained.

The SMBus router ensures that the operating system can communicate with individual batteries as well as the composite of batteries being discharged simultaneously. There is no requirement for the composite battery data to be generated within the electrical circuit (EC). Other alternatives, such as a private interface between the EC and the operating system that would allow a custom driver to calculate composite battery data, are allowed. The SMBus router also ensures that the operating system receives all alarms from the battery or batteries being charged or discharged.

7.2 Energy-Efficient Routing Protocol

A network is a collection of interconnected nodes. It can be wired, wireless, or wired cum wireless. A wireless ad hoc network has no fixed infrastructure. Wireless mobile networks and devices are becoming increasingly popular because they provide user access to information and communication anytime and anywhere. Since each node can work as a host as well as a router, there is no need for a separate router. The mobile ad hoc network has dynamic topology, a feature that helps it to change rapidly and unexpectedly.

Energy is a limiting factor in the case of ad hoc networks. Routing in these networks has some unique characteristics:

1. The energy of nodes is crucial and depends upon the battery, which has a limited power supply.
2. Nodes can move in an uncontrolled manner, so frequent route failures are possible.
3. Wireless channels have lower and more variable bandwidth compared to wired networks.

Energy-efficient design is very important in mobile ad hoc networks. MANETs consume more energy; as they have no fixed infrastructure, nodes should perform the operation of forwarding packets, along with routing. Therefore, traffic loads in MANETs are heavier than in any other wireless networks with fixed access points or base stations. For this reason, MANETs have more energy consumption. To make them more energy efficient, a trade-off is very necessary between different network performance criteria. To make a MANET more energy efficient, different protocols have been put forward, one of which is the energy-efficient medium access control (EE-MAC) protocol. The key idea behind this design is that most ad hoc networks are mostly data driven. Our goal in the EE-MAC protocol will be to reduce energy consumption without affecting its network performance. EE-MAC is based on the (IEEE) 802.11 standard, which is for wireless communication.

7.2.1 Proposed Energy-Efficient Medium Access Control Protocol

This protocol elects master nodes that keep awake and act as a network backbone to transfer data. The other (slave) nodes sleep to conserve energy and wake up periodically to communicate with the masters. To balance energy consumption, a rotation mechanism of masters and slaves is used. EE-MAC uses some features of power serving mode (PSM).

7.2.1.1 Design Criteria   The EE-MAC protocol is designed such that it contains enough master nodes to build the backbone of the network. Every node has at least one master nearby. Masters, collectively, are called the connected dominating set (CDS). The nodes in a network can fall under CDS or not. The nodes that fall under the CDS are masters, and the ones that do not are slaves. Slaves are free to sleep whenever they want, as they are not involved in the routing process.

Depending on the local information, the master nodes' algorithm is selected. To ensure that work is distributed fairly between master and slave, rotation is done according to an algorithm. This is very much needed because if rotation of master and slave is not done, a node will be overused and this may affect the network lifetime of an ad hoc network. If nodes are used as master and slave alternately, then we can have balance in energy consumption.

7.2.1.2 Features of EE-MAC   Some important features of EE-MAC are

1. **Entering sleep mode earlier.** One main drawback in PSM causes large energy consumption. Suppose that a node has some packets to send. It first sends an ATIM frame to the destination; in response to this, both source (transmitter) and destination (receiver) will be awake in that beacon interval, no matter how many packets need to be transmitted. The node will be awake unnecessarily, even if the single packet needs to be transmitted. This disadvantage is overcome in EE-MAC, where the information regarding the remaining number of packets is sent to the destination. This allows the destination to know when it has received all pending packets. When the source and destination have sent or received all their packets, they can enter sleep mode until the beginning of the next beacon interval.

2. **Priority processing of packets to slaves.** When a node has a certain number of packets to send, it first sends the packets to be sent for slave nodes. After transmitting the packets to the slave nodes, the packets destined for masters are transmitted. In brief, higher priority is given to slave nodes than to master nodes since this helps slave nodes to stay in sleep mode for a long time.

3. **Prolonging the sleep period for slaves.** The EE-MAC protocol is designed such that most of the packets are forwarded by masters, then slaves. To take advantage of this, each slave uses historical information to decide on its sleep time. Suppose that historical information has two consecutive beacon intervals and no packet is routed through slaves; these slaves decide to sleep for this time interval.

4. **Additional MAC layer control.** Nodes in an ad hoc network may move randomly as the topology selected will depend on the nearest possible route to the destination. Thus, to adapt quickly to network topology changes, nodes inform neighbors regarding their status (i.e., whether they are acting as masters or as slaves) by using the power-management bit in the MAC header. Because the MAC headers can be heard

anywhere in the network, including request to send and clear to send packets, this information will help neighbors to know each other's situation.

7.2.1.3 Performance   The network performance of an ad hoc network is evaluated by following metrics:

1. **Data packet delivery ratio.** The ratio of the number of packets generated at the sources to the number of packets received by the destination is called data packet delivery ratio. This signifies the throughput of the network. This metric is useful to measure any degradation in the network throughput.
2. **End-to-end delay.** This involves cumulative delay in the system as all possible delays caused by buffering, queuing, and retransmitting data packets, along with data propagation delay and transfer delays.
3. **Energy efficiency.** Energy efficiency is formulated as

    Energy efficiency = total number of bits transmitted/total energy consumed

where the total bits transmitted is calculated using application-layer data packets only, and total energy consumption is the sum of each node's energy consumption during the simulation time. The unit of energy consumption is bit per joule, and the greater the number of bits per joule, the better is the energy efficiency achieved.

7.3 Transmission Power-Management Schemes

The power-based connectivity definition is a new concept in wireless ad hoc networks. It attempts to improve end-to-end network throughput and the average power consumption. This is due to the fact that, as power gets higher and the connectivity range increases, each node will reach almost all other nodes in a single hop. However, since higher powers cause a higher interference level, more collisions occur, and hence there will be more transmission attempts. By reducing the transmission power levels at each node such that the node can directly connect to only a small subset of the network, the interference zones are considerably reduced.

Various routing algorithms have been proposed for wireless ad hoc networks in the literature. Those algorithms are mainly focused on establishing routes and maintaining these routes under frequent and unpredictable connectivity changes. The implicit assumption in most of the earlier work is that nodes' transmitted powers are fixed. To the best of our knowledge, there is no prior work that proposes the concept of mobile ad hoc nodes using different transmit powers. It is evident that this approach is restricted to ad hoc networks of relatively low mobility patterns. If the nodes are highly mobile, the power-management algorithm might fail to cope with the fast and sudden changes due to fading and interference conditions. We propose a power-management scheme that can be used in conjunction with traditional table-driven routing protocols, with possibly minor modifications. The performance measures are taken to be the end-to-end network throughput and the average power consumption.

7.3.1 *Power Management of Ad Hoc Networks*

An ad hoc network is wireless communication, which has the advantage of allowing untethered communication. This implies reliance on portable power sources such as batteries. However, due to the slow advancement in battery technology, battery power continues to be a constrained resource, so power management in wireless networks remains an important issue.

Various techniques, both in hardware and software, have been proposed to reduce energy consumption. Power management in ad hoc networks is a more difficult problem for two reasons:

1. In ad hoc networks, a node can be both a data source/sink and a router that forwards data for other nodes and participates in high-level routing and control protocols. Additionally, the roles of a particular node may change over time.
2. There is no centralized entity such as an access point to control and maintain the power-management mode of each node in the network, buffer data, and wake up sleeping nodes. Therefore, power management in ad hoc networks must be done in a distributed and cooperative fashion.

Power management in ad hoc networks spans all layers of the communication protocol stack. Each layer has access to different types of information about the communication in the network and thus uses different mechanisms for power management. The MAC layer does power management using local information, while the network layer can take a more global approach based on topology or traffic characteristics. We consider power-management approaches that save energy by turning off the radios of nodes in the network. Other energy conservation mechanisms such as topology control and power-controlled MAC protocols are considered orthogonal and the benefits can be combined.

Similarly to ad hoc routing protocols, power-management schemes range from proactive to reactive. The extreme of proactive can be defined as *always on* (i.e. all nodes are active all the time) and the extremity of reactive can be defined as *always off* (i.e., all nodes are in power-saving mode by default) (see Figure 7.7). Given the dynamic nature of ad hoc networks, there needs to be a balance between proactiveness, which generally provides more efficient communication, and reactiveness, which generally provides better power saving. Other techniques to reduce power consumption are shown in Table 7.2.

7.3.2 Basic Idea of the Power Cost Calculate Balance (PCCB) Routing Protocol

In ad hoc networks, designing an energy-efficient routing protocol is critical since nodes are power constrained. The PCCB (power cost calculate balance) routing protocol is proposed. The basic idea of PCCB is to utilize the "local multicast" mechanism and the energy boundary to select the route. Simulations show that the PCCB routing protocol can optimize power utilization and improve transmitting ratio. A PCCB network has a better performance than an AODV

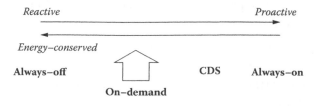

Figure 7.7 Design space of power-management schemes.

Table 7.2 Techniques to Reduce Power Consumption

PROTOCOL LAYER	POWER-CONSERVATION TECHNIQUES
Data-link layer	Avoid unnecessary retransmission. Avoid collision in channel access whenever possible. Put "receive" in standby mode whenever possible. Use or allocate contiguous slots for transmission and reception whenever possible. Turn radio off (sleep) when not transmitting or receiving.
Network layer	Consider route relaying load. Consider battery life in route selection. Reduce frequency of sending control message. Optimize size of control headers. Have efficient route reconfiguration techniques.
Transport layer	Avoid repeated retransmissions. Handle packet loss in a localized manner. Use power-efficient error control schemes.
Application layer	Adopt an adaptive mobile quality of service (QoS) framework. Move power-intensive computation from a mobile host to the base station. Use proxies for mobile clients. Proxies can be designed to make applications adapt to power or bandwidth constraints. Proxies can intelligently cache frequently used information, suppress video transmission and allow audio, and employ a variety of methods to conserve power.

network. The PCCB routing protocol is feasible for the energy-constrained character of mobile ad hoc networks.

7.3.2.1 Routing Process of the PCCB Routing Protocol

7.3.2.1.1 Protocol Assumption The following assumptions are made to simplify the model: (1) A node *can* get the value of its current energy, and (2) the links are bidirectional.

7.3.2.1.2 Route Discovery PCCB is a source-initiated on-demand routing protocol. As a result, nodes that are not on a selected path do not maintain routing information or participate in routing table exchanges. The PCCB routing protocol uses the following fields with route table entry:

1. Destination node address
2. Destination sequence number (guarantees the loop freedom of all routes toward that node)
3. Valid destination sequence number *flag*
4. Power boundary (the minimum energy of all nodes in the route)
5. Hop count (the number of hops needed to reach a destination)
6. Next hop
7. Lifetime (the expiration or deletion time of the route)

The route discovery of the PCCB is as follows:

1. If there is no direct valid path between source and the destination, the source node initiates a path discovery process to locate other nodes. The source node disseminates a route request (RREQ) to its neighbors. The RREQ includes information such as destination Internet protocol (IP) address, destination sequence number, power boundary, hop count, lifetime, etc. If no sequence number is known, the unknown sequence number flag must be set. The power boundary is equal to the source's energy; it will forward the packet if it matches some conditions.

2. When a node receives the RREQ from its neighbors, it first increments the hop count value in the RREQ by one, to account for the new hop through the intermediate node if the packet should not be discarded. The originator sequence number contained in the RREQ must be compared to the corresponding destination sequence number in the route table entry. If the originator sequence number of the RREQ is not less than the existing value, the node compares the power boundary contained in the RREQ to its current energy to get the minimum, and then it updates the power boundary of the RREQ with the minimum, which is the latest power boundary of this route.

3. Once the RREQ has arrived at the destination node or an intermediate node with an active route to the destination, the destination or intermediate node generates a route reply (RREP) packet and unicasts it back to the neighbor from which it received the RREQ. If the generating node is an intermediate node, it has an active route to the destination, the destination sequence number in the node's existing route table entry for the destination is not less than the destination sequence number of the RREQ, and the "destination-only" flag is not set. If the generating node is the destination itself, it must update its own sequence number to the maximum of its current sequence number and the destination sequence number in the RREQ packet immediately before it originates an RREP in response to an RREQ. The destination node places its (perhaps newly incremented) sequence number into the

destination sequence number field of the RREP and enters the value zero in the hop count field of the RREP. When generating an RREP message, a node wipes the destination IP address, originator sequence number, and power boundary from the RREQ message into the corresponding fields in the RREP message.

4. When a node receives the RREP from its neighbors, it first increments the hop count value in the RREP by one. As the RREP is forwarded back along the reverse path, the hop count field is incremented by one at each hop. Thus, when the RREP reaches the source, the hop count represents the distance, in hops, of the destination node from the source node. The originator sequence number contained in the RREP must be compared to the corresponding destination sequence number in the route table entry. If the originator sequence number of the RREP is not less than the existing value, the node compares the power boundary contained in the RREP to its current energy to get the minimum, and then it updates the power boundary of the RREP with the minimum, which is the latest power boundary of this route.

Note: If the sequence number in the routing table is marked as invalid in the route table entry or the destination sequence number in the RREP is greater than the node's copy of the destination sequence number, the intermediate node creates a new entry with the destination sequence number of the RREP and marks the destination sequence number as valid. The power boundary field in the route table entry is set to the power boundary contained in the RREP.

Note: If the originator sequence number contained in the RREP is equal to the existing destination sequence number in the node's route table, the power boundary of the RREP must be compared to the corresponding power boundary in the route table entry. If the power boundary contained in the RREP is greater than the node's copy of the power boundary, the power boundary in the entry is set to the value of the power boundary in the RREP. The next hop in the route entry is assigned to be the node from which the RREP is received, which is indicated by the source IP address field in the IP header. The current node can subsequently use this route to forward data packets to the destination.

7.3.2.1.3 Route Maintenance   According to the ad hoc on-demand distance vector (AODV) routing protocol, a node uses a HELLO message, which is a periodic local broadcast by a node to inform each mobile node in its neighborhood to maintain the local connectivity. If a node does not receive the HELLO message for a specified interval of time, called HELLO-loss HELLO-interval (which will be in milliseconds), the node should assume that the link to this neighbor is currently lost. When this scenario occurs, the node should send a route error (RERR) message to all precursors indicating which link has failed. Then the source initiates another route search process to find a new path to the destination or start the local repair.

7.3.3 Analysis of the PCCB Routing Protocol

Since nodes that are not on a selected path do not maintain routing information or participate in routing table exchanges, the PCCB routing protocol is a pure on-demand routing protocol. PCCB allows mobile nodes to obtain routes quickly for new destinations and respond to link breakages and changes in the network topology in a timely manner. When a link breaks, PCCB causes the affected set of nodes to be notified so that they are able to invalidate the routes using the lost link. PCCB uses power boundary as a selection criterion. If there are two routes to a destination, it is up to the requesting node to select a route that has a greater power boundary. The first priority to choose the route in PCCB is to select the shortest path. Then power boundary metrics are considered.

When the energy is almost exhausted, the operating system (OS) and basic input–output system (BIOS) will take actions in preparation for power down, which needs more power. The main advantage of the maximum power-boundary route is that it can reduce the additional information operations and conserve energy.

7.3.4 MAC Protocol

A medium access control (MAC) protocol decides when competing nodes may access the shared medium (i.e., the radio channel) and tries to ensure that no two nodes are interfering with each other's transmissions. In the unfortunate event of a collision, a MAC protocol

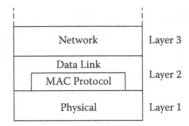

Figure 7.8   Network protocol stack (MAC layer).

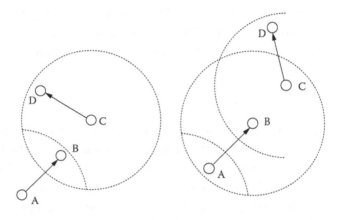

Figure 7.9   Hidden-terminal problem (left) and exposed-terminal problem (right).

may deal with it through some contention resolution algorithm. The MAC protocol layer is shown in Figure 7.8. The two main problems or collisions faced are the exposed-terminal problem and the hidden-terminal problem. These problems are depicted in Figure 7.9. In the hidden-terminal problem (left), communication is established, although it possibly interferes with packet reception of neighboring stations. In the exposed-terminal problem (right), no communication is established because of an ongoing communication between a pair of neighbor stations, although their communication will not be impaired. The IEEE 802.11 MAC protocol deals with this and other problems and fulfills the requirements appropriately as mentioned before. The basic MAC protocol is similar to the 802.3 MAC protocol.

7.3.5 *Power Saving*

In the IEEE 802.11 power-saving mechanism specified for ad hoc networks, each station is synchronized by beacon frame transmission at the beginning of a period called the beacon interval. The stations that successfully exchange control packets during the ATIM window start to transmit data immediately after the ATIM window expires. When network load is increased, the stations waste energy in contention and retransmission operations, which are significant sources of energy consumption and channel utilization degradation. A new beacon interval structure removes the ATIM window and divides the rest of the beacon interval into equal interval time slots. The distributed coordination function (DCF) with different interframe space and priority-based time-slot occupation is used to provide quality of service, increase time in the doze state, and decrease the number of collisions.

7.3.6 Timing Synchronization Function

The timing synchronization function (TSF) is specified in the IEEE 802.11 wireless local area network (WLAN) standard to fulfill timing synchronization among users. A TSF keeps the timers for all stations in the same basic service set (BSS) synchronized. All stations should maintain a local TSF timer. Each mobile host maintains a TSF timer with a modulus 2^{64} counting in increments of microseconds. The TSF is based on a 1 MHz clock and "ticks" in microseconds. On a commercial level, industry vendors assume the 802.11 TSF's synchronization to be within 25 µs.

Timing synchronization is achieved by stations periodically exchanging timing information through beacon frames. Each station in an independent basic service set (IBSS) should adopt a received timing if it is later than the station's own TSF timer. All stations in the IBSS adopt a common value—a beacon period—that defines the length of beacon intervals or periods. This value, established by the station that initiates the IBSS, defines a series of target beacon transmission times (TBTTs) exactly a beacon period time unit apart. Time zero is defined to be a TBTT.

7.3.7 Power-Saving Function

The IEEE 802.11 standard specifies two medium access methods: DCF (distribution coordination function), for a fully distributed protocol, and PCF (point coordination function) or a centralized protocol.

DCF is fundamental access method of the IEEE 802.11 MAC and should be implemented in all stations. On the other hand, PCF is an option access method, which is only usable in an infrastructure network. IEEE 802.11 defines two power saving mechanisms depending on the configuration of network: infrastructure network (BSS) and ad hoc network (IBSS). The following subsections briefly describe the DCF protocol and power-saving mechanism in IBSS.

In the IEEE 802.11 distribution coordination function, a station with pending packets to transmit has to monitor the channel status before transmitting the data. If a channel is idle for a period, called distributed interframe space (DIFS), the station uniformly chose back-off time in range [0, cw], where CW is the size of the contention window. The value CW is set to CWmin at the first transmitting attempt. The back-off time decreases when the channel is sensed to be idle, holds the same value when the channel is busy, and decreases again when the channel is sensed to be idle for DIFS. When the back-off time becomes zero, the station transmits the RTS packet. On reception of the RTS packet, the destination station responds with the CTS packet to the transmitting station. The transmitting station starts to send data after receiving the CTS packet.

When the destination station receives a data packet successfully, it sends an ACK (acknowledgment) packet to the transmitting station to inform it of the successful reception. Any station in the network that overhears the RTS and CTS packets can update a network allocation vector (NAV) by information indicating the length of the packet to be transmitted in both packets. Note that before the transmit CTS packet, data packet, and ACK packet, the station waits for a short interframe space (SIFS) instead of DIFS. On the other hand, if back-off time in two or more stations reaches zero at the same time, a collision occurs when they send the RTS packet. In this case, the destination station will not receive the RTS packet and will not respond with the CTS packet. The transmitting station can detect the collision by absence of the CTS packet. Every time a collision is detected, the contention window is doubled to reduce the probability of collision again, up to the maximum value CWmax.

The power-saving mechanism for an IBSS divided time into a constant interval named the beacon interval. At the start of the beacon interval, every wireless host has to stay awake for a fixed period, called

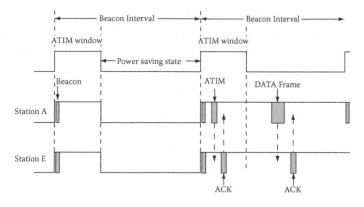

Figure 7.10   Example of transmitting to a power-saving station in an IBSS network.

the ATIM (ad hoc traffic indication message) window. As described in the IEEE 802.11 standard, every station in the network is assumed to be fully connected and synchronized. Because of characteristics of the network, there is no unlimited power station playing a role to provide time synchronization. Every wireless host attempts to broadcast the beacon frame using a CSMA/CA mechanism at the start of the beacon interval to obtain synchronization. After receiving the beacon frame, the wireless host adjusts its local time to the time stamp of the beacon frame to synchronize with each other.

During the ATIM window, the wireless host that buffers data for a station in power-saving mode announces an ATIM frame. On receiving the ATIM frame, the power-saving host must send ATIM-ACK back. After successfully transmitting the ATIM frame, both wireless hosts will stay in awake states for the whole beacon interval. After the end of the ATIM window, buffered data will be transmitted to a destination by the normal distribution coordination function protocol. At this point, a wireless station that has no buffered packet can change to doze mode if it does not receive an ATIM frame during the ATIM window. Figure 7.10 illustrates the example of transmitting data in an ad hoc network to a power-saving station.

7.3.8 Power-Saving Potential

The effectiveness of the power-saving mechanism depends heavily on the values selected for the beacon and ATIM intervals, as well as the offered load. If the ATIM window is too short, not enough traffic can be announced during the window—possibly not even enough to utilize the entire beacon interval. If the ATIM window is too long, not only are stations required to spend more time awake, but also more traffic may be announced than can be sent in the remainder of the beacon interval. If the beacon interval is too short, the overhead of the sleep–wake cycle, beaconing, and traffic announcements will be high. If the beacon interval is too long, many stations may attempt to announce traffic at each ATIM window, such that many destinations will need to remain awake after the ATIM window.

The key criterion is naturally the amount of power savings, but factors such as increased latency, decreased throughput, and unequal distribution of power consumption must also be taken into account. It is important to note that the percentage of time spent in the sleep state is only an indication of the actual energy savings, which will be reduced by the costs of the wake–sleep transition, beaconing, and ATIM traffic, all of which increase as the beacon interval decreases.

7.4 Transmission Power Control

Transmit power control is important in wireless ad hoc networks for at least two reasons: (1) It can impact on battery life, and (2) it can impact the traffic-carrying capacity of the network.

For the first point, note that there is no need for N1 in Figure 7.11 to broadcast at 30 mW to send a packet to the neighboring N2, since N2 is within range even at 1 mW. Thus, it can save on battery power. For the second point, suppose that in the same figure, N3 also wishes to broadcast a packet at the same time to N4 at 1 mW. If N1 broadcasts at 1 mW to N2, then both transmissions can be successfully received simultaneously, since N2 is not in the range of its interferer N3 (for its reception from N1), and N4 is not in the range of its interferer N1. However, if N1 broadcasts at 30 mW, then that interferes with N4's reception from N3, so only one packet, from N1 to N2, is successfully transmitted. Thus, power control can enhance the traffic-carrying capacity.

One wants an adaptive choice of power level by nodes in the network, which is implementable in a distributed asynchronous fashion by the nodes participating in the network. The next issue that arises is where in the layered hierarchy the power control for ad hoc networks fits. The difficulty is that it infringes on several layers.

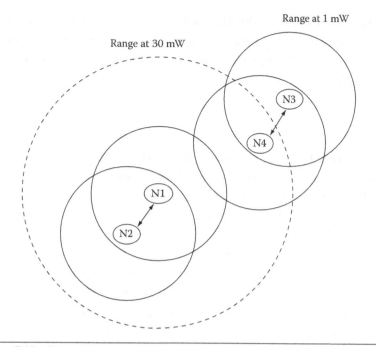

Figure 7.11 The need for power control.

Clearly, power control impacts the physical layer due to the need for maintaining link quality. However, power control also impacts the network layer, as shown in Figure 7.11(a). If all nodes are transmitting at 1 mW, then the route from node N1 to node N5 is N1 → N2 → N3 → N4 → N5. However, if they all transmit at 30 mW, then one can choose the route N1 → N3 → N5. In addition, power control also impacts on the transport layer. In Figure 7.11(b), every time node N1 transmits at high power to node N2, it causes interference at N3 to the packets from N4. Thus, there is a loss of several such packets on the link from N4 to N3. This impacts the congestion control algorithm regulating the flow from source N4 to destination N5 via the intermediate relay node N3. The need for power control is thus obvious.

7.4.1 Adapting Transmission Power to the Channel State

Depending on the channel state, transmission power can also be adapted on short time scales. Therefore, if the channel quality is estimated as better, a lower transmission power is assigned, resulting in less interference and power consumption by the amplifier.

On the other hand, if the channel quality is estimated as worse, a higher transmission power is assigned or the transmission process might be stopped temporarily until the communication channel is good again.

7.4.2 MAC Techniques

Power-saving techniques existing at the MAC layer consist primarily of sleep-scheduling protocols. The basic principle behind all sleep-scheduling protocols is that lots of power is wasted listening on the radio channel while there is nothing there to receive. Sleep schedulers are used to duty cycle a radio between its on and off power states in order to reduce the effects of this *idle listening*. They are used to wake up a radio whenever it expects to transmit or receive packets and sleep otherwise. Other power-saving techniques at this layer include battery-aware MAC protocols (BAMAC) in which the decision of who should send next is based on the battery level of all surrounding nodes in the network. Battery-level information is piggybacked on each packet that is transmitted, and individual nodes base their decisions for sending on this information.

Sleep scheduling protocols can be broken up into two categories: synchronous and asynchronous. Synchronous sleep scheduling policies rely on clock synchronization between all nodes in a network, as seen in Figure 7.12. Senders and receivers are aware of when each other should be on and only send to one another during those time periods. They go to sleep otherwise.

Asynchronous sleep scheduling, on the other hand, does not rely on any clock synchronization between nodes whatsoever. Nodes can send and receive packets whenever they please, according to the MAC protocol in use. Figure 7.13 shows how two nodes running asynchronous sleep schedulers are able to communicate.

Nodes wake up and go to sleep periodically in the same way they do for synchronous sleep scheduling. Since there is no time synchronization, however, there must be a way to ensure that receiving nodes are awake to hear the transmissions coming in from other nodes. Normally, preamble bytes are sent by a packet in order to synchronize the starting point of the incoming data stream between the transmitter and receiver. With asynchronous sleep scheduling, a significant

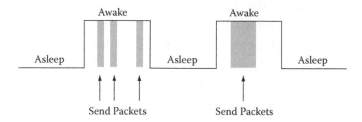

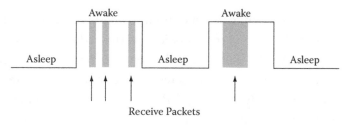

Figure 7.12   Synchronous sleep scheduler.

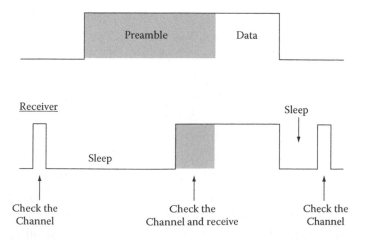

Figure 7.13   Asynchronous sleep scheduler.

number of extra preamble bytes are sent per packet in order to guarantee that a receiver has the chance to synchronize to it at some point. In the worst case, a packet will begin transmitting just as its receiver goes to sleep, and preamble bytes will have to be sent for a time equal to the receiver's sleep interval (plus a little more to allow for proper synchronization once it wakes up). Once the receiver wakes up, it synchronizes to these preamble bytes and remains on until it receives the packet.

7.4.3 *Logical Link Control*

The two most common techniques used to conserve energy at the link layer involve reducing the transmission overhead during the *automatic repeat request* (ARQ) and *forward error correction* (FEC) schemes. Both of these schemes are used to reduce the number of packet errors at a receiving node. By enabling ARQ, a router is able automatically to request the retransmission of a packet directly from its source without first requiring the receiver node to detect that a packet error has occurred. Results have shown that sometimes it is more energy efficient to transmit at a lower transmission power and have to send multiple ARQs than to send at a high transmission power and achieve better throughput. Integrating the use of FEC codes to reduce the number of retransmissions necessary at the lower transmission power can result in even more energy savings. Power management techniques exist that exploit these observations

Hybrid scheme: With ARQ, the system performs well in good channels, but not in poor ones, whereas FEC performs consistently across channel conditions due to a lack of feedback, though it has the capability of trading off more protection for increased computational cost. Therefore, the combined effect will create minimal energy consumed per useful bit over the code rate.

Adaptive frame sizing: A different frame size can have a dramatic impact on the behavior of the ARQ protocol. In a noisy channel, a smaller packet will be preferable despite the excess overhead, since the packet loss rates of the larger packets will dominate. Turning this around, an argument can be made for low-power packet sizing. If we are to accept a fixed, low user level bit rate, we can trade off reduced power transmissions for packet throughput.

Other power-management techniques existing at the link layer are based on some sort of packet-scheduling protocol. By scheduling multiple packet transmissions`1 to occur back to back (i.e., in a burst), it may be possible to reduce the overhead associated with sending each packet individually. Preamble bytes only need to be sent for the first packet in order to announce its presence on the radio channel, and all subsequent packets essentially piggyback this announcement. Packet-scheduling algorithms may also reduce the number of retransmissions necessary if a packet is only scheduled to be sent during a time when its destination is known to be able to receive packets. By reducing the number of retransmissions necessary, the overall power consumption is reduced as well.

Routing: In traditional cellular systems, the routing problem changes to the handover problem: Which access point should be selected to serve a mobile device? As a handover decision is usually based on the channel quality between mobile devices and access points, energy efficiency is at least implicitly considered in this process. From a system perspective, it could be conceivable that a handover decision that is energy conserving for a particular mobile device could be suboptimal when considering all mobiles together; this suboptimality could, for example, be due to a different interference situation in neighboring cells. The routing problem becomes much more difficult if multihop radio communication is considered. In such a multihop system, it is no longer clear which sequence of nodes should be traversed to reach a given destination.

Several optimization metrics can be introduced to support this choice. A number of routing protocols have been developed to meet the specific needs of such multihop networks—for example, proactive protocols like destination-sequenced distance vector (DSDV), which periodically sends route updates to learn all routes to a destination in the network; reactive protocols like dynamic source routing (DSR) and power-aware routing optimization (PARO), which start searching for the destination only if there is a packet to transmit; and hybrid schemes like AODV, FSR, and the temporally ordered routing algorithm (TORA). However, energy efficiency is not the prime target of these protocols.

More recently, energy efficiency has moved into focus, particularly motivated by the vision of wireless sensor networks. A frequently used concept is to assign routing and forwarding responsibilities to a node acting on behalf of a group of nodes (a "cluster"); routing and forwarding then only take place among these "cluster heads." The choice of cluster heads can be based on the availability of resources (battery capacity) and is rotated among several nodes in many approaches. Examples of such clustered protocols are the zone routing protocol (ZRP) and low-energy adaptive clustering hierarchy (LEACH). Additionally, some routing protocols take the physical location of nodes into account (e.g., geographical adaptive fidelity, or GAF). The challenge for all of these multihop routing protocols is the evaluation of the trade-off between energy savings by clever routing and the overhead required to obtain the routing information, particularly in the face of uncertainties induced by mobility, time-varying channels, and so forth.

7.5 AODV Protocol

7.5.1 Introduction

The ad hoc on-demand distance vector (AODV) routing protocol provides a method of routing in mobile ad hoc networks. This means that routes are only established when needed to reduce traffic overhead. AODV supports unicast, broadcast, and multicast without any further protocols. Link breakages can be locally repaired very efficiently.

7.5.2 Route Discovery

When a source has data to transmit to an unknown destination, it broadcasts a route request (RREQ) for that destination. At each intermediate node, when an RREQ is received, a route to the source is created. If the receiving node has not received this RREQ before, is not the destination, and does not have a current route to the destination, it rebroadcasts the RREQ. If the receiving node is the destination or has a current route to the destination, it generates a route reply (RREP). The RREP is unicast in a hop-by-hop fashion to the source. As the RREP propagates, each intermediate node creates a route to the destination. When the source receives the RREP, it records the route to the destination and can begin sending data. If multiple

RREPs are received by the source, the route with the shortest hop count is chosen.

As data flow from the source to the destination, each node along the route updates the timers associated with the routes to the source and destination, maintaining the routes in the routing table. If a route is not used for some period of time, the node removes the route from its routing table.

7.5.3 Route Maintenance

Route maintenance is done as follows: If data are flowing and a link break is detected, a route error (RERR) is sent to the source of the data in a hop-by-hop fashion. As the RERR propagates toward the source, each intermediate node invalidates routes to any unreachable destinations. When the source of the data receives the RERR, it invalidates the route and reinitiates route discovery if necessary.

7.6 Local Energy-Aware Routing Based on AODV (LEAR-AODV)

7.6.1 Introduction

The purpose of LEAR-AODV is to balance the energy consumption rates throughout the network. This is done by allowing the nodes to choose whether they will be part of a route or not. The choice is based on the remaining battery power that a node has. In other words, a node can chose to reduce its participation in data forwarding and therefore conserve power. The protocol incorporates a mechanism that is used to avoid shortage of forwarding nodes due to selfish behavior. To make all of this possible, the route discovery and route maintenance procedures of AODV are modified.

7.6.2 Route Discovery

During route discovery, when an intermediate node receives a route request packet, it first examines its remaining battery power. If it is less than some predetermined threshold, the route request packet is dropped and the node announces that by broadcasting an ADJUST_ Thr packet. This automatically means that the node will not forward the data packet on behalf of the source node that sent the route request. Otherwise, if the intermediate node has sufficient battery power, it retransmits the packet. To that end, it is guaranteed that the destination node will receive a route request along a route of nodes with sufficient battery power.

7.6.3 Route Maintenance

The route maintenance procedure in AODV is triggered by the unavailability of a hop along a source-destination route. An intermediate node that identifies a missing hop reports back to the source node and, as a result, a new route discovery is initiated. In the case of LEAR-AODV, a route maintenance procedure could be initiated by a node with decreasing battery power. Nodes of the network are continuously checking their remaining battery power. If it becomes lower than the threshold value as a result of an ongoing data transfer, the node issues a route maintenance packet to the source node indicating that it will no longer be a part of the corresponding route.

LEAR-AODV provides a mechanism for a real-time adjustment of the threshold battery power values. This is to avoid the situation in which route request packets do not reach the destination node due to low battery power of the intermediate nodes. In such a case, after an unsuccessful route request, the source node issues its following route request with an indication that the intermediate nodes must decrease the battery power threshold value.

7.7 Power-Aware Routing Based on AODV (PAR-AODV)

7.7.1 Introduction

PAR-AODV assigns costs to each hop that lies on a source-destination route, based on the residual battery power of each node. Using these costs, all available routes are evaluated. The protocol uses the route that minimizes the following function:

$$C(\pi, t) = \Sigma_{i \in \pi} C_i(t) \tag{7.1}$$

where

$$C_i(t) = \rho_i \left(\frac{F_i}{E_i(t)} \right) a \qquad (7.2)$$

and

ρ_i is the transmit power of node i

F_i is the full-charge battery capacity of the node i

E_t is the remaining battery capacity of node i in time t

a is a positive weighting factor

7.7.2 Route Discovery

During route discovery, prior to the transmission of a route request packet, each intermediate node calculates its link cost using Equation (7.2) and adds it to the header of the packet. Thus, when the destination node receives the route request packet, it sends a route reply back to the source that contains the overall cost of the route. The source node selects the route that offers the lowest cost.

Additional *compute-cost* packets could be sent by the intermediate nodes in case they receive route request packets with a lower link cost than that currently in use. The compute_cost packets are sent to the destination node, which then informs the source node of the new, more cost-effective route, using a route reply.

7.7.3 Route Maintenance

The route maintenance in PAR-AODV is the same as in LEAR-AODV. When any intermediate node has a lower battery level than its threshold value, any request is simply dropped.

7.8 Lifetime Prediction Routing Based on AODV (LPR-AODV)

7.8.1 Introduction

The last of the power-aware routing protocols proposed is LPR-AODV. It routes traffic through paths with a predicted long lifetime. As in the case of PAR-AODV, the protocol assigns a cost to each link. The cost used by LPR-AODV is related to the battery lifetime of a node. The chosen route is the one that maximizes the function

$$max_\pi \left(T_\pi \left(t \right) \right) = max_\pi \left(min_{i \in \pi} \left(T_i(t) \right) \right) \tag{7.3}$$

where $T_\pi(t)$ is the lifetime of path π and $T_i(t)$ is the predicted lifetime of node i in path π. The battery lifetime prediction of a node is based on its past activities. A good indication of the amount of traffic crossing the node is achieved by keeping a log of recent data-routing operations. Every time the node sends a data packet, it records its residual battery energy $Ei(t)$ at the given time instance t. The node also logs its residual energy $Ei(t)$ at time instance i when exactly N packets are sent/forwarded.

7.8.2 Route Discovery

Similarly to LEAR-AODV, each intermediate node calculates its costs in terms of predicted lifetime T_i using the following formulas:

$$T_i = \frac{E_i(t)}{discharge\_rate_i(t)} \tag{7.4}$$

where

$$Discharge_{rate_i}(t) = \frac{E_i(t') - E_i(t)}{t - t'} \tag{7.5}$$

where $E_i(t)$ is the remaining energy of node i at time t.

The estimated node cost is inserted by the intermediate nodes in the header of the propagated route request packet. On reception of a route request, the destination node issues a route reply that contains the overall route cost. If an intermediate node receives a route request packet with lower cost, the destination node is informed by a *compute_lifetime* packet. Thereafter, the destination node informs the source node about the new route with a route reply packet.

7.8.3 Route Maintenance

As in the first algorithms, route maintenance is needed when a node becomes out of direct range of a sending node or when there is a change in its predicted lifetime. In the first case (node mobility), the mechanism is the same as in AODV. In the second case, the node sends an RERR back to the source even when the predicted life-time goes below a threshold level δ ($Ti(t) = \delta$). This route error message forces the source to initiate route discovery again. This decision depends only on the remaining battery capacity of the current node and its discharge rate. Hence, it is a local decision. However, the same problem as in LEAR-AODV can occur. If the condition $Ti(t) = \delta$ is satisfied for all the nodes, the source will not receive a single reply message even though a path between the source and the destination exists. To prevent this, we use the same mechanisms used in LEAR-AODV described earlier.

The three algorithms are simulated and compared to the unmodified AODV protocol under two different scenarios: fixed and mobile. The improved network lifetime is studied in terms of

- The time taken for K nodes to die
- The time taken for the first node to die
- The time taken for all nodes to die

In the static case, the best performance is observed for LPR-AODV, where the first node to switch off due to exhausted power resources under AODV routing appears 3244 s before a node malfunctions under LPR-AODV. This protocol outperforms the others by taking into account the battery discharge rates in addition to residual battery capacity.

In the mobile case, the LPR-AODV protocol once again offers the best performance in terms of network lifetime extension. All three algorithms outperform the unmodified AODV algorithms under all mobility instances with an average network life extension of 1033 s at node speed of 4 m/s. As in the case of EADSR, with increasing mobility, the energy consumption performance of the modified versions of AODV converges to that of the original protocol.

Problems

7.1 Why is energy management an important factor in ad hoc networks?

7.2 Explain how energy-management schemes are classified in ad hoc networks.

7.3 Explain different battery management schemes for ad hoc networks.

7.4 Briefly explain the overview of battery technologies used for ad hoc networks.

7.5 Discuss the principles of battery discharge.

7.6 Discuss the impact of discharge characteristics on battery capacity.

7.7 Explain how a smart battery system can be implemented with an example.

7.8 Give an overview of battery-driven system design.

7.9 Describe the features and design criteria of the EE-MAC protocol.

7.10 Explain the analysis of the PCCB routing protocol.

7.11 Give a brief description of the timing synchronization function.

7.12 Explain the power-saving function used in power-saving mechanisms.

7.13 Describe logical link control for power transmission.

7.14 Give an overview of the AODV protocol.

7.15 Describe local energy-aware routing based on AODV.

7.16 Explain power-aware routing based on AODV.

7.17 Describe lifetime prediction routing based on AODV.

References

1. Lahiri, K., A. Raghunathan, S. Dey, and D. Panigrahi. 2002. Battery-driven system design: A new frontier in low power design. *ASP-DAC '02: Proceedings of the 2002 Asia and South Pacific Design Automation Conference,* p. 261.

2. Prabhu, B. J., A. Chockalingam, and V. Sharma. 2002. Performance analysis of battery power management schemes in wireless mobile devices. *IEEE Wireless Communications and Networking Conference* 2:825–831.

3. Nie, J., and Z. Zhou. 2004. An energy based power-aware routing protocol in ad hoc networks. *IEEE International Symposium on Communications and Information Technology* 1:280–285.

4. Senouci, S. M., and G. Pujolle. 2004. Energy efficient routing in wireless ad hoc networks. *IEEE International Conference on Communications* 7:4057–4061.

5. Langendoen, K., and G. Halkes. Energy-efficient medium access control, Delft University of Technology, Faculty of Electrical Engineering, Mathematics and Computer Science, Delft, the Netherlands, *The Embedded Systems Handbook.*

6. Jayashree, S., and C. Siva Ram Murthy. 2007. A taxonomy of energy management protocols for ad hoc wireless networks. *IEEE Communications Magazine* 45:104–110.
7. Benchmarq Microelectronics Inc., Duracell Inc., Energizer Power Systems, Intel Corporation, Linear Technology Corporation, Maxim Integrated Products, Mitsubishi Electric Corporation, National Semiconductor Corporation, Toshiba Battery Co., Varta Batterie, AG. Smart battery system manager specification. 1988.
8. Senouci, S. M., and G. Pujolle. 2004. Energy efficient routing in wireless ad hoc networks. *IEEE International Conference on Communications* 7:4057–4061.
9. Chen, M., and Gabriel A. Rincon-Mora. 2006. Accurate electrical battery model capable of predicting runtime and I–V performance. *IEEE Transactions on Energy Conversion* 21 (2): 504–511.

8

MOBILITY MODELS FOR MULTIHOP WIRELESS NETWORKS

8.1 Introduction

In the simulation environment to test a new protocol for an ad hoc network, a mobility model needs to be used for representing the movements of the mobile nodes (mobile nodes) that will properly utilize the given protocol. These models play a vital role in determining the protocol performance in MANET. Hence, it is essential to study and analyze the effects of various mobility models on the performance of the MANET protocols. Using a mobility model we try to mimic the real movements of the nodes using a particular networking scenario. The mobility models were primarily designed for representing the movement pattern of mobile users, and how their location, velocity, and acceleration change over time. The mobile nodes are inherently dynamic in nature and different kinds of mobility models have been proposed in order to capture various features of mobility of the nodes.

8.2 Mobility Models

In the simulation environment we use the mobility models to describe the movements of the mobile nodes. They define the position, speed, and acceleration of the nodes at every moment in the simulation scenario. Based on the dependence of the movements, these mobility models can be classified in

- Entity mobility models (independent movements)
- Group mobility models (dependent movements)

We have basically two types of mobility models that are used in the simulation of networks: traces and synthetic models. Traces mobility models are those mobility patterns that are observed in real life systems. In the case of a large number of nodes and a long observation period, they provide accurate information. New ad hoc network environments can not be easily modeled if traces have not yet been created. In such a scenario, synthetic models can be used. Synthetic models try to represent the behaviors of mobile nodes realistically without the use of traces.

In the case of a synthetic model, the models need to be designed in such a way that they should change the speed and direction in a reasonable time slot. For example, we would not want the mobile nodes to travel in straight lines at constant speeds throughout the course of the entire simulation that does not match the behaviour of the real nodes. There are seven different synthetic entity mobility models for ad hoc networks:

- The random walk mobility model (including its many derivatives) is a simple mobility model based on random directions and speeds.
- The random waypoint mobility model includes pause times between changes in destination and speed.
- The boundless simulation area mobility model converts a 2D rectangular simulation area into a torus-shaped simulation area.
- The random direction mobility model forces the mobile nodes to travel to the edge of the simulation area before changing direction and speed.
- The probabilistic version of the random walk mobility model utilizes a set of probabilities to determine the next position of a mobile node in such models.
- The Gauss–Markov mobility model uses one tuning parameter to vary the degree of random nodes in the mobility pattern.
- The city section mobility model is a simulation area that represents streets of a city.

The first two models —the random walk mobility model and the random waypoint mobility model—are commonly used by researchers.

8.2.1 Random Walk Mobility Model

Einstein, in 1926, mathematically described the first random walk mobility model (Figure 8.1). Many entities in nature move in extremely unpredictable ways, and the random walk mobility model was developed to ape this unpredictable movement. According to this mobility model, a mobile node moves from its current location to a new location by randomly choosing a new direction and speed. The new speed and direction are both chosen from pre-defined ranges. Each of the movements in this model takes place in either a constant time interval t or a constant distance traveled d, after which a new direction and speed are calculated. When a node reaches a simulation boundary, it "bounces" off the border of the simulation area with an angle that is determined by the incoming direction. The mobile node then follows this new path.

Many variations of the random walk mobility Model have been developed including the 1-D, 2-D, 3-D, and d-D walks. This characteristic ensures that the random walk represents a mobility model that tests the movements of nodes around their starting points, without worrying about them wandering away never to return.

The random walk mobility model does not keep or use the past locations and speed values; hence, it can be called a memory-less model. The current speed and direction of a mobile node are not

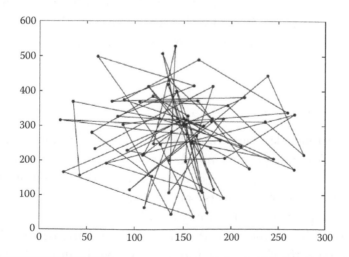

Figure 8.1 Traveling pattern of an mobile node using the 2-D random walk mobility model (time).

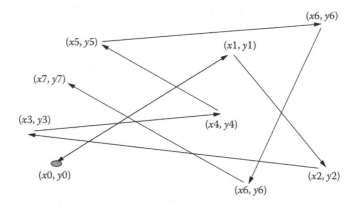

Figure 8.2   Static nature of random walk mobility model.

dependent on its past speed and direction. Because of this, it can generate unrealistic movements like sharp turns and sudden stops.

Movement pattern of nodes in simulation is a random roaming pattern restricted to a small segment of the simulation area. Certain simulation studies based on this mobility model set the specified time to one clock tick or the specified distance to one step. From Figure 8.2 we observe the static nature obtained in the random walk mobility model if the mobile node is permitted to move 10 steps (not one) before changing direction. It may also be observed that the mobile node does not roam far from its initial position. The random walk mobility model is used widely in simulation. For further simplification of the model, all the mobile nodes can be assigned the same speed.

8.2.2 Random Waypoint

This model includes pause times between changes in direction and/ or speed. According to this model, a mobile node begins by residing at one location for a specific period of time. At the expiration of this time duration, the node selects a random destination in the simulation area and a speed that is uniformly distributed between [*minspeed* and *maxspeed*]. The mobile node then moves toward the newly selected destination at the chosen speed. After reaching the destination it has to again wait for the specified pause time before repeating the process.

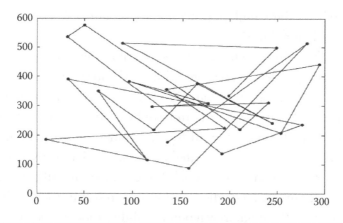

Figure 8.3 Traveling pattern of a mobile node using the random waypoint mobility model.

Figure 8.3 shows an example traveling pattern of a mobile node that uses the random waypoint mobility model. The node starts at a randomly chosen point or position (133, 180) whereas the speed of the mobile node is uniformly chosen between 0 and 10 m/s. It can also be observed that the movement pattern of a mobile node that uses the random waypoint mobility model is similar to the random walk mobility model if pause time is zero and [*minspeed, maxspeed*] = [*speedmin, speedmax*]. This mobility model has been widely used by different researchers. Moreover, the model is sometimes simplified. For example, the random waypoint mobility model is used without pause times. In most of the performance evaluations using the random waypoint mobility model, the mobile nodes are initially distributed randomly around the entire simulation area. This initial random distribution of the nodes is not the same in which nodes distribute themselves while moving.

The average mobile node neighbor percentage is the cumulative percentage of total mobile nodes that are a given mobile node's neighbor. For example, if the network contains 50 mobile nodes and a node has 5 neighbors, then the node's current neighbor percentage is 10%. A node is considered as a neighbor of another node if it is within the node's transmission range. As can be seen from Figure 8.3, there is high variability during the first 600 seconds of simulation time. This high variability in average mobile node neighbor percentage will cause a high variability in performance results when the simulation results are calculated from relatively short simulation duration.

In the following, we present three possible solutions to avoid this initialization problem is presented. First, the locations of the mobile nodes can be saved after a simulation has executed long enough to be past this initial high variability, and can be used as the initial starting point of the mobile nodes in all future simulations. Second, initially the mobile nodes can be distributed in a manner that maps to a distribution more common to the model. For example, initially placing the mobile nodes in a triangle distribution may result in the distribution of the nodes in the random waypoint mobility model more accurately than distributing the mobile nodes randomly in the simulation area. Lastly, the initial 1000 seconds of simulation time produced by the random waypoint mobility model in each simulation trial needs to be discarded. (Not considering the first 1000 seconds of simulation time eliminates the initialization problem even if the mobile nodes move slowly. In other words, we can discard fewer seconds of simulation time for faster moving mobile nodes.) This approach has an added advantage over the first solution proposed. Specifically, a random initial configuration for each simulation is ensured by this simple solution. In this case, a performance evaluation is based on the random waypoint mobility model, and appropriate parameters need to be evaluated. For example, a multicast protocol for ad hoc networks can be evaluated by using the random waypoint mobility model.

8.2.3 The Random Direction Mobility Model

In order to overcome density waves in the average number of neighbors produced by the random waypoint mobility model, the random direction mobility model was created. A density wave is considered as the clustering of nodes in one part of the simulation area. For the random waypoint mobility model, the clustering occurs near the center of the simulation region. In the case of random waypoint mobility model, the probability of a mobile node choosing a new destination—which is located in the center of the simulation region or a destination that requires traveling through the middle of the simulation region—is high.

8.2.4 A Boundless Simulation Area

In the boundless simulation area mobility model, we find a relationship between the previous direction of travel and velocity of a mobile node with its current direction of travel and velocity. In order to describe a mobile node's velocity v and its direction θ, a velocity vector v = (v,θ) is specified; the location of the mobile node is represented as $(x; y)$. Changes in both the velocity vector and the position occur at every Δt time steps according to the following formulas:

$$v(t + \Delta t) = \min[\max(v(t) + \Delta v, 0), V_{max}];$$
$$\theta(t + \Delta t) = \theta(t) + \Delta\theta;$$
$$x(t + \Delta t) = x(t) + v(t) * \cos\theta(t);$$
$$y(t + \Delta t) = y(t) + v(t) * \sin\theta(t);$$

Where V_{max} referes to the maximum velocity defined in the simulation, Δv refers to the change in velocity, which is uniformly distributed between $[-V_{max} * \Delta t; A_{max} * \Delta t]$, A_{max} is the maximum acceleration of a given mobile node, $\Delta\theta$ is the change in direction, which is uniformly distributed between $[-\alpha * \Delta t; \alpha * \Delta t]$, and α is the maximum angular change in the direction in which a mobile node is traveling.

The boundless simulation area mobility model is also different in terms of how the boundary of a simulation area is handled. In case of all the mobility models previously discussed, mobile nodes reflect off or stop moving once they reach a simulation boundary. But in the boundless simulation area mobility model, the mobile nodes that reach one end of the simulation region continue their movement and reappear on the opposite side of the simulation region. This technique creates a torus-shaped simulation area allowing mobile nodes to travel unobstructed (see Figures 8.4 and 8.5).

8.2.5 Gauss–Markov

The Gauss–Markov mobility model was primarily designed for adapting to different levels of randomness through one tuning parameter. Initially, a current speed and direction are assigned to each mobile node. At fixed time intervals, n, movement of each node occurs by updating the speed and direction of the node. The value of speed and direction at the nth instance is determined depending upon their

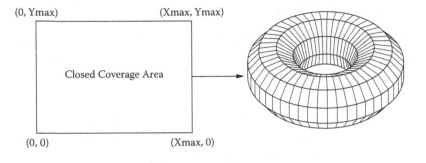

(0, Ymax)   (Xmax, Ymax)

Closed Coverage Area

(0, 0)   (Xmax, 0)

Figure 8.4 Rectangular simulation area mapped to a torus in the boundless simulation area mobility model.

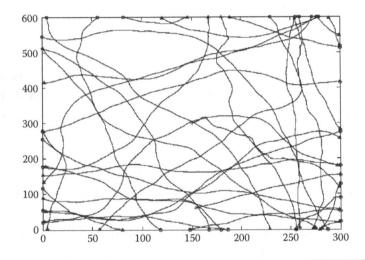

Figure 8.5 Traveling pattern of a mobile node using the boundless simulation area mobility model.

respective values at the $(n-1)^{st}$ instance and a random variable according to the following equations

$$s_n = \alpha\, s_{n-1} + (1-\alpha)\bar{s} + \sqrt{\left(1-\alpha^2\right)}\, s_{x_{n-1}}$$

$$d_n = \alpha\, d_{n-1} + (1-\alpha)\bar{d} + \sqrt{\left(1-\alpha^2\right)}\, d_{x_{n-1}}$$

where s_n and d_n specify the value of the new speed and direction of the mobile node at time interval n; α, where $0 \leq \alpha \leq 1$, refers to the tuning parameter used to vary the randomness; s and d are constants representing the mean value of speed and direction as $n \to \infty$; and $s_{x_{n-1}}$ and $d_{x_{n-1}}$ are random variables from a Gaussian distribution. Totally random values (or Brownian motion) are obtained by setting $\alpha = 0$ and linear motion is obtained by setting $\alpha = 1$. Varying the value of α between 0 and 1, intermediate levels of randomness are obtained. At each time interval, the next location is determined depending on the current location, speed, and direction of movement. Specifically, at time interval n, an mobile node's position can be obtained by the equations

$$x_n = x_{n-1} + s_{n-1} \cos \theta\, d_{n-1}$$

$$y_n = y_{n-1} + s_{n-1} \cos \theta\, d_{n-1}$$

where (x_n, y_n) and (x_{n-1}, y_{n-1}) denote the x and y coordinates of the mobile node's position at the nth and $(n-1)^{st}$ time intervals, respectively, and s_{n-1} and d_{n-1} denote the speed and direction of the mobile node, respectively, at the $(n-1)^{st}$ time interval. In order to ascertain that a mobile node does not remain near an edge of the grid for a long duration, the mobile nodes are forced away from an edge when they reach within a certain distance of the edge. This is achieved through the modification of the mean direction variable d in the above direction equation. For example, when a mobile node is present near the right edge of the simulation area, the value of d is changed to 180 degrees. Consequently, the mobile node's new direction is away from the right edge of the simulation area.

8.2.6 A Probabilistic Version of Random Walk

In the case of Chiang's mobility model, a probability matrix is used to determine the position of a particular mobile node in the next time step, which is represented by three different states for position x and three different states for position y. State 0 denotes the current (x or y) position of a mobile node, state 1 denotes the mobile node's previous (x or y) position, and state 2 denotes the mobile node's next position when the mobile node continues to move in the same direction. The probability matrix used is

$$P = \begin{bmatrix} P(0,0) & P(0,1) & P(0,2) \\ P(1,0) & P(1,1) & P(1,2) \\ P(2,0) & P(2,1) & P(2,2) \end{bmatrix}$$

where each entry $P(a,b)$ denotes the probability for a mobile node to go from state a to state b. These values are used for updating both the mobile node's x and y positions. In Chiang's simulator, an average speed is preset for each node for their random movement. The following matrix contains the values Chiang used to calculate x and y movements:

$$P = \begin{bmatrix} 0 & 0.5 & 0.5 \\ 0.3 & 0.7 & 0 \\ 0.3 & 0 & 0.7 \end{bmatrix}$$

With these values, a mobile node may take a step in any of the four possible directions (i.e., north, south, east, or west) during its movement (i.e., no pause time). Moreover, the probability of the mobile node moving in the same direction is higher than the probability of the mobile node that changes directions. Also, the values defined prohibit movement between the previous and next positions without passing through the current location. A probabilistic rather than purely random movements is produced by this implementation, which may result in more realistic behaviors. For example, as people complete their day-to day activities they tend to continue their movement in a semi-constant forward direction. Rarely do they suddenly turn around to retrace their steps. Also, random steps are almost never taken hoping that they may eventually wind up somewhere relevant to their tasks. However, the task of selecting appropriate values of $P(a;b)$ may be difficult, if not impossible, for an individual simulation scenario unless traces are available for a given movement scenario.

8.2.7 City Section Mobility Model

In the case of the city section mobility model, the simulation scenario represents a section of a city where the ad hoc network exists. The type of city being simulated determines the streets and speed limits on the streets. For example, in the downtown area of the city the streets may form a grid and a high-speed highway near the border of the simulation area to represent a loop around the city. Under this mobility model every mobile node starts the simulation at a defined point on some street. It then randomly chooses the next destination, also represented by a point on some street. The movement algorithm from the current position to the new destination finds a path that will produce the shortest travel time between the two points. Moreover, safe driving characteristics including speed limit and a minimum distance allowed between any two mobile nodes exist. After reaching the destination, the mobile node has to again pause for a specified time and before randomly choosing another destination (i.e., a point on some street). After this the process is repeated.

8.3 Limitations of the Random Waypoint Model and Other Random Models

The design characteristics of the random waypoint model and its variants try to mimic the movement of mobile nodes in a simplified way. Their wide acceptability is mainly due the simplicity of implementation and analysis. However, this also results in certain limitations such that they may not adequately capture certain mobility characteristics of some realistic scenarios, including temporal dependency, spatial dependency and geographic restriction:

- **Temporal dependency of velocity.** In random waypoint and other random models, the velocity of mobile node is considered as a memory-less random process, (i.e., the velocity of current movement is not dependent on the velocity of the previous movement). This may result in some extreme mobility behavior, such as sudden stop, sudden acceleration, and sharp turn, in the trace generated by the random waypoint model. However, in most of the real life scenarios, the speed of vehicles and pedestrians accelerate incrementally. Moreover, the change in direction is also smooth.
- **Spatial dependency of velocity.** The mobile node is basically considered as an entity with independent movement in the case of the random waypoint and other random models. This kind of mobility model is categorized as an entity mobility model. However, in some scenarios, including battlefield communication and museum touring, the movement pattern of a mobile node may vary under the influence of certain

specific "leader" node in its neighborhood. Consequently, the mobility of various nodes is indeed correlated.

*   **Geographic restrictions of movement.** In the random waypoint and other random models, the movement of the mobile nodes is free from any restrictions. However, in many realistic cases, especially for the urban area applications, the movement of a mobile node may be constrained by obstacles, buildings, streets, or freeways. The random waypoint model and its variants fail to represent these constraints.

8.3.1 *Mobility Models with Temporal Dependency*

The physical laws of acceleration, velocity, and rate of change of direction constrain the mobility of a node. Hence, there may be a dependency between the current velocity of a mobile node and its previous velocity. Thus, a correlation exists between the velocities of a single node at different time slots. This mobility characteristic is known as the temporal dependency of velocity. However, due to their memoryless nature of the random walk model, the random waypoint model and other variants are not capable of capturing this temporal dependency behavior. As a result, various mobility models considering temporal dependency are proposed.

8.3.2 *Mobility Models with Spatial Dependency*

In the random waypoint model and other random models, the movement of a particular mobile node is independent of other nodes. For example the location, speed, and movement direction of mobile node are not affected by other nodes in the neighborhood. These models are not capable of capturing many realistic scenarios of mobility. For example, on a freeway in order to avoid collision, the speed of a vehicle should not exceed the speed of the vehicle in front of it. In addition to this, in some specific MANET applications such as disaster relief and battlefield, team collaboration among users is necessary and the users are likely to follow the team leader. Hence, the mobility of a mobile node could be influenced by the mobility of other neighboring nodes. As a correlation exists in space between the velocities of different nodes, we call this characteristic as the spatial dependency of velocity.

8.3.3 Mobility Models with Geographic Restriction

In the random waypoint model, the nodes are allowed to move freely and randomly anywhere within the simulation area. In contrast, in most real life scenarios, it can be observed that a node's movement is subject to the environment. Specifically, the motions of vehicles are bounded to the freeways or local streets in the urban area, and on campus the pedestrians may be blocked by the buildings and other obstacles. This may result in a pseudo-random movement pattern of the nodes on predefined pathways in the simulation area. In some recent works this characteristic has been addressed by integrating the paths and obstacles into mobility models. This kind of mobility model is known as a mobility model with geographic restriction.

8.3.3.1 Pathway Mobility Model   In order to integrate geographic constraints into the mobility model it is necessary to restrict the node movement to the pathways in the map. The map is predefined in the simulation area. Tian, Hahner, and Becker et al. utilize a random graph to model the map of city which can be either randomly generated or carefully defined depending on certain map of a real city. The buildings of the city are represented by the vertices of the graph whereas the edges model the streets and freeways between those buildings.

Initially, the nodes are placed randomly on the edges of the graph. Then, for each node, a random next destination is chosen. The node then moves toward the selected destination following the shortest path along the edges. Upon arrival, the node pauses for T time and again chooses a new destination for the next movement. This procedure is repeated for the entire simulation duration. Unlike the random waypoint model where the nodes can move freely, the mobile nodes in this model are restricted to travel on the pathways. However, due to the random selection of the destination for each phase, a certain level of randomness still exists for this model. Consequently, in the graph-based mobility model, the nodes travel following a pseudo-random pattern on the pathways. Similarly, in the case of the freeway mobility model and Manhattan mobility model, the movement of mobile nodes is also restricted to the pathway in the simulation region.

8.3.3.2 Obstacle Mobility Model   The obstacles in the simulation field also play a major role as geographic constraint in mobility models. For avoiding the obstacles on the path of movement, it is necessary for the mobile node to change its trajectory. Therefore, the movement behavior of mobile nodes is affected by the obstacles. Additionally, the obstacles also affect the way radio propagates. For example, for the indoor environment, typically, it is not possible for the radio system to propagate the signal through obstacles without severe attenuation. In the case of the outdoor environment, the radio is also subject to the radio shadowing effect. Therefore, while integrating obstacles into the mobility model, we must carefully consider both its effect on node mobility and on radio propagation. Johansson, Larsson, and Hedman et al. developed three 'realistic' mobility scenarios to represent the movement of mobile users in real life, including

1. The conference scenario consists of 50 people attending a conference. The majority of them are static and a small number of people are moving with low mobility.
2. In the event coverage scenario, a group of highly mobile people or vehicles are modeled. Here the mobile nodes frequently change their positions.
3. In the disaster relief scenarios, some nodes move very fast whereas others move very slowly.

For all the above mobility scenarios, obstacles in the form of rectangular boxes are randomly placed on the simulation region. It is required for the mobile node to select a proper movement trajectory for avoiding such obstacles. Additionally, when the radio propagates through an obstacle, it is assumed that the signal is fully absorbed by the obstacle. More specifically, if an obstacle resides within two nodes, the link between these nodes is considered as broken until one of them moves out of the shadowed area of the other. Thus, under these effects, the three proposed mobility scenarios seem to differ from the commonly used random waypoint model. Jardosh, Belding-Royer, and Almeroth et al. also investigated the impact of obstacles on mobility modeling in detail. After taking into consideration the effects of obstacles into the mobility model, both the movement trajectories and the radio propagation of mobile nodes are somehow restricted.

In the simulation region, a number of obstacles are placed to model the buildings. Thus, based on the locations of the building or obstacles, a Voronoi graph is computed to construct the pathways. The mobile nodes are restricted to move only on the pathways that interconnect the buildings. The pathways constructed by the Voronoi graph are equidistant from the nearby buildings. This observation is consistent with the real scenario where the pathways tend to lie half-way in-between the adjacent buildings. Additionally, in this model, the nodes (e.g., students on campus) are allowed to enter and exit buildings.

After the construction of the pathway graph, the movements of mobile nodes are restricted on the pathways. This causes the mobile nodes to travel in a semi-definitive (i.e., pseudo random) manner. When the mobile node moves towards its randomly selected new destination on the pathway graph, it follows the shortest path through the predefined pathway graph. In the Voronoi diagram this shortest path is calculated by Dijikstra's algorithm.

8.3.3.3 Group Mobility Models There are certain situations where it is necessary to model the behavior of mobile nodes as they move together. For example, in a battlefield scenario, a group of soldiers may be assigned the task of searching a particular plot of land in order to destroy land mines, capture enemy attackers, or simply work together in a cooperative manner to accomplish a common goal. For modeling such situations, a group mobility model is highly essential to simulate this cooperative feature.

8.4 Summary

After examining the various mobility models, we have conducted a survey of the mobility modeling and analysis techniques systematically. In this chapter, a detailed discussion and study have been conducted not only for the random waypoint model and its variants, but also for several other mobility models with unique characteristics such as temporal dependency, spatial dependency, or geographic restriction. We believe that the set of mobility models considered here reasonably represent the state-of-the-art research and technology in this field. After the detailed study of these mobility models, it can be observed that the mobility models may have various properties and may exhibit different mobility features. Consequently, these mobility models may behave differently and influence the protocol performance in various ways. Hence, for proper evaluation of ad hoc protocol performance, it is necessary to use a rich set of mobility models instead of single the random waypoint model. Every model in the set has its own unique and specific mobility characteristics. As a result, a proper method for correctly choosing a suitable set of mobility models is needed.

The performance of an ad hoc network protocol can vary significantly under the influence of different mobility models. Even under the same mobility model, the performance of an ad hoc network protocol can vary significantly when used with different parameters.

The selection of a mobility model may require a data traffic pattern that has significant effect on the protocol performance. For instance, when a group mobility model is simulated, then protocol evaluation should be carried out with a portion of the traffic local to the group. Intra-group communication will have significant changes on a protocol's performance, compared to the same mobility scenarios and all inter-group communication.

For proper performance evaluation of an ad hoc network protocol we must use the mobility model that most closely matches the expected real-world scenario. In fact, the development of the ad hoc network protocol can be aided significantly by the anticipated real-world scenario. However, considering the development of ad hoc networks as a relatively new field of research, it may be noted that it is still not clear what a realistic model is for a given scenario.

Problems

8.1   Why are mobility models needed for a wireless environment?

8.2   Give the classification of mobility models.

8.3   Justify the need for characterizations of mobility.

8.4   Discuss the classification of mobility patterns.

8.5   Explain the column node model with a suitable example.

8.6   Describe the random waypoint mobility model with an example.

8.7   What are the limitations of the random waypoint mobility model?

8.8   What are temporal dependency models? Explain.

8.9   Describe the Gauss–Markov mobility model with a neat diagram.

8.10 What are spatial dependency models? Explain with an example.

8.11 Explain the geographic restriction model with an illustration.

References

1. Davies, V. 2000. Evaluating mobility models within an ad hoc network. Master's thesis, Colorado School of Mines.
2. Tian, Hahner, Becker, et al.
3. Johansson, Larsson, Hedman, et al.
4. Jardosh, Belding-Royer, Almeroth, et al.

Bibliography

Aguayo, D., J. Bricket, S. Biswas, G. Judd, and R. Morris. 2004. Link-level measurements from an 802.11b mesh network. *ACM SIGCOMM Computer Communication Review* 34:121–132.

Bar-Noy, A., I. Kessler, and M. Sidi. 1994. Mobile users: To update or not to update? *Proceedings of the Joint Conference of the IEEE Computer and Communications Societies (INFOCOM)*, pp. 570–576.

Basagni, S., I. Chlamtac, V. R. Syrotiuk, and B.A. Woodward. 1998. A distance routing effect algorithm for mobility (DREAM). *Proceedings of the ACM/IEEE International Conference on Mobile Computing and Networking (MOBICOM)*, pp. 76–84.

Bettstetter, C., G. Resta, and P. Santi. 2003. The node distribution of the random waypoint mobility model for wireless ad hoc networks. *IEEE Transactions on Mobile Computing* 2 (3): 257–269.

Boleng, J. 2001. Normalizing mobility characteristics and enabling adaptive protocols for ad hoc networks. *Proceedings of the Local and Metropolitan Area Networks Workshop (LANMAN)*, pp. 9–12.

Broch, J., D. Maltz, D. Johnson, Y. Hu, and J. Jetcheva. 1998. Multi-hop wireless ad hoc network routing protocols. *Proceedings of the ACM/IEEE International Conference on Mobile Computing and Networking (MOBICOM)*, pp. 85–97.

Camp, T., J. Boleng, and V. Davies. 2002. A survey of mobility models for ad hoc network research. *Wireless Communications and Mobile Computing (WCMC):* Special issue on mobile ad hoc networking: research, trends and applications 2:483–502.

Camp, T., J. Boleng, B. Williams, L. Wilcox, and W. Navidi. 2002. Performance evaluation of two location based routing protocols. *Proceedings of the Joint Conference of the IEEE Computer and Communications Societies (INFOCOM)* 3:1678–1687.

Chiang, C. 1998. Wireless network multicasting. PhD thesis, University of California, Los Angeles.

Chiang, C., and M. Gerla. 1998. On-demand multicast in mobile wireless networks. *Proceedings of the IEEE International Conference on Network Protocols (ICNP)*, pp. 262–270.

Fitzek, F., L. Badia, M. Zorzi, G. Schulte, P. Seeling, and T. Henderson. 2003. Mobility and stability evaluation in wireless multihop networks using multiplayer games. ACM NETGAMES '03, Redwood City, CA.

Garcia-Luna-Aceves, J. J., and E. L. Madrga. 1999. A multicast routing protocol for ad-hoc networks. *Proceedings of the Joint Conference of the IEEE Computer and Communications Societies (INFOCOM)*, pp. 784–792.

Garcia-Luna-Aceves, J. J., and M. Spohn. 1999. Source-tree routing in wireless networks. *Proceedings of the 7th International Conference on Network Protocols (ICNP)*, pp. 273–282.

Gerharz, M., C. de Waal, M. Frank, and P. Martini. 2002. Link stability in mobile wireless ad hoc networks. *Proceedings of the IEEE Conference on Local Computer Networks (LCN)*, Tampa, FL, pp. 230–239.

Haas, Z. 1997. A new routing protocol for reconfigurable wireless networks. *Proceedings of the IEEE International Conference on Universal Personal Communications (ICUPC)*, pp. 562–565.

Hong, X., M. Gerla, G. Pei, and C.-C. Chiang. 1999. A group mobility model for ad hoc wireless networks. Proceedings of ACM/IEEE MSWiM'99, Seattle, WA, pp. 53–60.

Hui, P., A. Chaintreau, J. Scott, R. Gass, J. Crowcroft, and C. Diot. 2005. Pocket switched networks and human mobility in conference environments. *Proceedings of the ACM SIGCOMM 2005 Workshop on Delay-Tolerant Networking (WDTN)*, Philadelphia, PA, pp. 244–251.

IETF Mobile Ad Hoc Networking (MANET) Working Group. 2004. http://www.ietf.org/html.charters/manet-charter.html.

Jetcheva, J. G., Y.-C. Hu, S. PalChaudhuri, A. K. Saha, and D. B. Johnson. 2003. Design and evaluation of a metropolitan area multitier wireless ad hoc network architecture. *Proceedings of the IEEE Workshop on Mobile Computing Systems and Applications (WMCSA)*, Pittsburgh, PA, pp. 32–43.

Johnson, D. B., D. A. Maltz, Y.-C. Hu, and J. G. Jetcheva. 2002. The dynamic source routing protocol for mobile ad hoc networks (DSR).

Kotz, D., and K. Essien. 2005. Analysis of a campus-wide wireless network. *Wireless Networks* 11:115–133.

Lenders, V., J. Wagner, and M. May. 2006. Measurements from an 802.11b mobile ad hoc network. *Proceedings of the IEEE WoWMoM workshop on advanced experimental activities on wireless networks and systems (EXPONWIRELESS)*, Niagara-Falls/Buffalo, NY, pp. 519–524.

McDonald, A. B., and T. Znati. 1999. A path availability model for wireless ad hoc networks. *Proceedings of the IEEE Wireless Communications and Networking Conference (WCNC)* 1:35–40.

McNett, M., and G. M. Voelker. 2005. Access and mobility of wireless PDA users. *ACM SIGMOBILE Mobile Computing and Communications Review* 9:40–55.

Petrak, L., O. Landsiedel, and K. Wehrle. 2005. Framework for evaluation of networked mobile games. *NetGames '05*, Hawthorne, New York, pp. 1–7.

Sadagopan, N., F. Bai, B. Krishnamachari, and A. Helmy. 2003. PATHS: Analysis of PATH duration statistics and their impact on reactive MANET routing protocols. *Proceedings of the Fourth ACM International Symposium on Mobile Ad Hoc Networking and Computing*, Annapolis, MD, pp. 245–256.

Su, J., A. Chin, A. Popivanova, A. Goel, and E. de Lara. 2004. User mobility for opportunistic ad hoc networking. *Proceedings of the IEEE Workshop on Mobile Computing Systems and Applications (WMCSA)*, English Lake District, UK, pp. 41–50.

Tuduce, C., and T. Gross. 2005. A mobility model based on WLAN traces and its validation. *Proceedings of INFOCOM*, Miami, FL, 1:664–674.

9
CROSS-LAYER
DESIGN ISSUES

9.1 Introduction

As wireless communications and networking fast occupy center stage in research and development activity in the area of communication networks, the suitability of one of the foundations of networking—the layered protocol architecture—is coming under close scrutiny from the research community. It is repeatedly argued that although layered architectures have served well for wired networks, they are not suitable for wireless networks. To illustrate this point, researchers usually present what they call a cross-layer design proposal.

Due to lack of coordination among the layers, the performance of layered architecture posed peculiar challenges in the area of ad hoc wireless networks. To overcome the limitations of layered architecture, cross-layer design was proposed; the idea here is to maintain the functionalities associated with the original layers but to allow coordination, interaction, and joint optimization of protocols crossing different layers. Compared to strict layered architecture, performance of cross-layer architecture is better with reference to protocol design done by dependency between protocol layers. Unlike layering, protocols at the different layers are designed independently because of several analyses of cross-layer design. From a literature survey of many IEEE papers, we came to know that there are many cross-layer design proposals that are weak in performance and implementation.

9.2 A Definition of Cross-Layer Design

The open system interconnection (OSI) model is a seven-layered architecture recommended by the International Standards Organization

275

(ISO). It divides the overall networking task into layers and introduces a hierarchy of services to be provided by the individual layers. Each layer is associated with certain set of protocols providing communication among the corresponding layer between the computers and forbids direct communication between nonadjacent layers. In layered architecture, protocols are designed with respect to the rules of reference architecture, this means that the higher layer protocol only makes use of the services at the lower layers and is not concerned about the details of how the services are provided. Instead, protocols can be designed by violating the rules of reference architecture of OSI by allowing direct communication between protocols at nonadjacent layers. Such violations of a layered architecture by communicating with nonadjacent layers are called cross-layer design.

9.3 Cross-Layer Design Principle

Figure 9.1 shows the traditional OSI layered architecture, the major mechanism in the success of the Internet. The OSI model is organized and divided into layers; each layer is built on top of the one below, and each layer should fulfill a limited and well defined purpose. Each

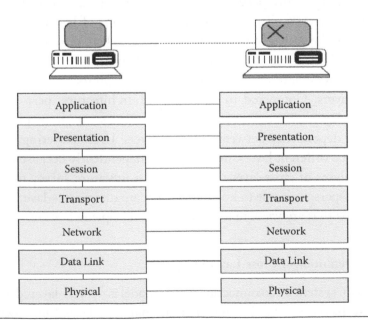

Figure 9.1   The layered OSI architecture.

lower layer provides services to the respective layers above and it encapsulates the data by providing an abstract interface for its services. This has led to rapid growth in the development of a number of applications that drive the Internet.

The dynamic nature of wireless networks makes layered architecture suffers from suboptimality and inflexibility because each layer has insufficient information about the network. It does not allow sharing of information among the layers dynamically. In wireless networks, layers must coordinate and adapt to the changes. This is the motivation behind the cross-layer design in ad hoc wireless networks.

9.3.1 General Motivations for Cross-Layer Design

As discussed in the previous section, the presence of wireless links in the existing network motivates designers to violate the layered architecture principles because of unique problems created by wireless links. In turn, a wireless medium needs some of the requirements for communication of those facilities not supported by layered architecture. For instance, multiple packet reception is made by the physical layer at the same time.

Figure 9.2 shows a block diagram of a new cross-layer design framework with information exchange between the different layers. At the link layer (lower layer), adaptive modulations are used to maximize the link rates under varying channel conditions. This extends the achievable capacity region of the network. Each point of this region indicates a possible assignment of the different link capacities. Based on link-state information (service), the medium access control (MAC; immediate upper layer) selects one point of the capacity region by assigning time slots, codes, or frequency bands to each of the links. The MAC layer operates jointly with the network layer to determine the set of network flows that minimize congestion.

Solutions for capacity assignments and network flows are exchanged iteratively between two middle layers that constitute the core of the cross-layer framework (network and MAC layer). At the transport layer, congestion control and retransmission of packets takes place. Finally, the application layer determines the most efficient encoding rate.

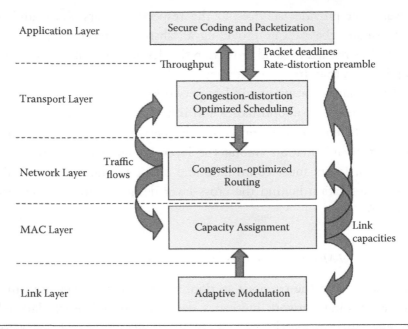

Figure 9.2   Cross-layer design framework with information exchange between different layers.

9.4 Proposals Involving Cross-Layer Design

In this section we will study different architecture violations in order to improve the performance of the network in the cross-layer design. Following are the basic ways of violating the layered architecture:

- Creation of new interfaces
- Merging of adjacent layers
- Design coupling without new interfaces
- Vertical calibration across layers

9.4.1 *Creation of New Communication Interfaces*

Communication interfaces define the way that service primitives of a protocol provide services to its upper layer and use services from its lower layers. The interfaces are also called service access points (SAPs). The newly created communication interfaces are used for information sharing between the layers at runtime. Creation of new communication interfaces is a violation of layered architecture because this type of information sharing is not supported by layered architecture.

Depending on the direction of information flow, this category can be further divided into three subcategories:

- Upward information flow
- Downward information flow
- Back and forth information flow

9.4.1.1 Upward Information Flow   In the upward information flow, the flow of information from the lower layer to a higher layer protocol at runtime results in the creation of a new SAP from lower to higher layer, as shown in Figure. 9.3(a). For instance, if the end-to-end transmission control protocol (TCP) path contains a wireless link, errors on the wireless link can trick the TCP sender into making erroneous inferences about the loss of a packet in the network, and as a result the performance deteriorates. Creating interfaces from the lower layers to the transport layer to enable explicit notifications helps such situations. For example, the loss of packet information from the TCP receiver to the transport layer at the TCP sender can explicitly tell the TCP sender to retransmit the packets if there is a loss of packets in the network.

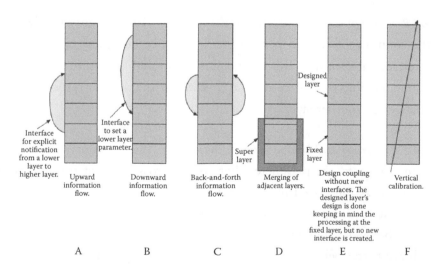

Figure 9.3   Illustrating the different kinds of cross-layer design proposals. The rectangular boxes represent the protocol layers

9.4.1.2 Downward Information Flow   In the downward information flow, flow of information from the higher layer to the lower layer protocol at runtime results in the creation of a new SAP from higher to lower layer, as shown in Figure. 9.3(b). Here, the purpose is to provide hints to the lower layers about how the higher layer data processed.

9.4.1.3 Back and Forth Information Flow   Here, two layers are involved at the same time to perform different tasks. At runtime this is an iterative loop between the two layers, with the information flowing back and forth, as shown in Figure 9.3(c). Because of the complementary new interfaces, this clearly violates the layered architecture.

9.4.2 Merging of Adjacent Layers

Another category of cross-layer design is merging of adjacent layers. In this category, merging of services provided by the essential layers is called a superlayer. This design does not require any creation of new communication interfaces. The created superlayer can communicate with the rest of the stack using interfaces that were already in the original layered architecture.

9.4.2.1 Design Coupling without New Interfaces   This category involves the coupling of two or more layers without creating any extra new interfaces at runtime. This category is illustrated in Figure 9.3(e). it is not possible to replace one layer without making corresponding changes to another layer.

9.4.2.2 Vertical Calibration across Layers   In this category, as the name suggests, the parameter that spans across layers at runtime is adjusted, as illustrated in Figure 9.3(f). Performance of application layers depends upon the involvement of various parameters on the layers below it. Therefore, it is believed that joint tuning can help to achieve better performance than the individual layer parameters.

Parameters (as would happen had the protocols been designed independently) can achieve.

9.5 Proposals for Implementing Cross-Layer Interactions

Depending on cross-layer interaction, implementation can be divided into three categories:

- Direct communication between layers
- A shared database across layers
- Completely new abstractions

9.5.1 Direct Communication between Layers

As the name suggests, runtime information sharing allows layers to communicate directly with each other, as shown in Figure 9.4(a). Here, communication is transparent; this means making the variables at one layer visible to the other layer at runtime. Here, there are many ways in which the layers can communicate with one another via protocol header or extra information header.

9.5.2 A Shared Database across Layers

As the name suggests, in this category a common database is available that can be accessed by all the layers, as illustrated in Figure 9.4(b). The common database is like a new layer providing the service of storage/retrieval of information to/from all the layers. This category is well suited to vertical calibration as discussed in earlier section. The main challenge in this category is to design new interactions between the different layers and a common database.

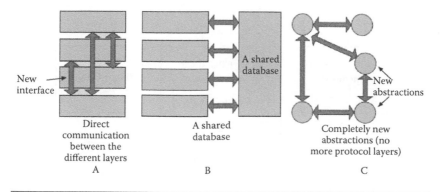

Figure 9.4 Proposals for architectural blueprints for wireless communications.

9.5.3 Completely New Abstractions

As this category suggests, this is a completely new abstraction with no more protocol layers, which we depict schematically in Figure 9.4(c). This category offers flexibility during design as well as at runtime due to rich interactions between protocols.

9.6 Cross-Layer Design: Is It Worth Applying It?

As discussed in earlier sections, layered architecture is not supportive and flexible in data sharing among layers for the dynamic nature of nodes in wireless networks. In order to overcome these limitations, here we discuss why cross-layer design should be approached in a careful manner. Although the layered architecture may not be optimal in the theoretical sense, the performance enhancement that it guarantees is the longevity of the system and low implementation costs. Examples of such layered architecture are found in the following subsections.

9.6.1 The von Neumann Architecture

The von Neumann architecture (Figure 9.5) is the heart of most computer systems. It includes the independent functional units: the memory unit, control unit, arithmetic and logical unit, and input–output unit. The architecture makes it possible that hardware and software for the computer systems may be developed independently. This is one of the major reasons behind the rapid proliferation of computer systems.

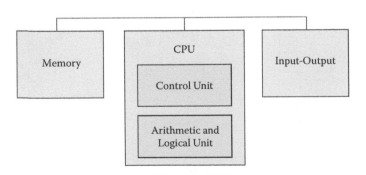

Figure 9.5   The von Neumann architecture for computer systems.

9.6.2 Source-Channel Separation and Digital System Architecture

In his seminal work on information theory, Shannon proved that the layers of source compression (source coding) and coding for reliable transmission over wireless channel (channel coding) may be implemented separately and independently. This implied that each new source of information (and the associated source encoder/decoder) may simply reuse the existing channel encoders/decoders, thus simplifying the implementation. In the same way, new (more efficient) channel encoders/decoders may be designed without worrying about the sources that would be using the channel. This architecture has fueled rapid development and proliferation of digital communication systems (Figure 9.6).

9.6.3 The OSI Architecture for Networking

The OSI architecture (Figure 9.1) and its impact on development and proliferation of computer networks have been sufficiently discussed in this chapter. The historical evidence thus seems to support layered design architectures. We now discuss some of the possible technical and economic disadvantages of cross-layer design.

9.7 Pitfalls of the Cross-Layer Design Approach

9.7.1 Cost of Development

The main aim of the cross-layer design is that it should be flexible according to the network state such that it optimizes the performance of applications. If the demands of two applications or environments of two instances of wireless network are radically different, then each of them would require a separate set of protocols. Thus, the cross-layers would have to be handcrafted for each application and network scenario.

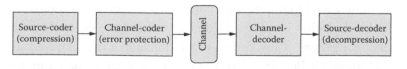

Figure 9.6 Source-channel separation and the architecture of digital communication systems.

For example, consider two networks, N1 and N2. Suppose that in N1 the nodes are battery operated, while in N2 the nodes have an infinite source of power (e.g., connected to an electrical socket or operated on solar power). Therefore, in N1 one of the objectives of the optimization would be to consume minimum energy, while in N2 this would not be valid. Although the protocol set of N1 would work fine in N2 also, this again would lead to suboptimal performance in N2, to counter which the cross-layer design was developed in the first place. Thus, handcrafting of protocols for each application and network would lead to high deployment costs and delays.

9.7.2 Performance versus Longevity

The most popular argument in favor of sharing information among layers is that it leads to optimal performance. However, such a gain in performance is of a short-sighted nature because the technologies at each layer change rapidly. Thus, every change of technology, the nature of information that is shared, and the actions that are taken would need to be changed. This is against the principle of longevity, which is considered an essential feature of any design. If we include the weight of longevity and cost while evaluating the overall performance of architecture, then the layered architecture may as well outperform the architectures that makes aggressive use of the cross-layer approach.

9.7.3 Interaction and Unintended Consequences

The layered architecture allows limited and controlled interactions between the layers such that the job of designing or modifying a protocol at any level is simplified. A cross-layered approach leads to many dependencies between various layers. The designer of a new protocol in a cross-layer system has to understand and take into account the interaction of various layers. In spite of a good understanding, a new protocol may lead to unintended consequences due to the presence of multiple adaptation loops. Such interactions need to be studied using dependency graphs. Thus, design of a protocol in a cross-layered stack is much more challenging than the task of designing such a protocol in a layered stack.

9.7.4 Stability

As already mentioned, a cross-layer design leads to several adaptation loops. The complex interaction of these loops may endanger the stability of the system. Although cross-layer design offers tremendous opportunities, at the same time it has several critical disadvantages that may hinder the proliferation of wireless networks. The cross-layer approach should thus be used with caution. Much functionality, like transmit power control and channel state estimation, is typical of wireless networks and requires a cross-layer approach. In cases where the cross-layer design is necessary, it should be ensured that the implementation is not very aggressive (i.e., does not rely too much on information exchanged among several layers). The placement of functions and the nature of the information exchanged between the layers must be kept to a minimum and must be critically analyzed.

9.8 Performance Objectives

In the case of wireless ad hoc networks, performance objectives can be divided into two categories:

- Power-based performance objectives that maximize network lifetime. A typical example is a sensor network. The traffic requirement is low, and the main goal is to maintain a network in operation as long as possible.
- Rate-based performance objectives. For these, the goal is to maximize flow rates. Typical examples are wireless LANs, networks of computer peripherals, or home appliances.

Here, we considered the wireless networks with best-effort traffic, focused on rate-based performance metrics, and analyzed the well-known rate-based performance metrics originally defined in the wired networking context: rate maximization, proportional fairness and max–min fairness, and some of their modifications. The three most frequently used design objectives for wireless networks are *maximizing total capacity, max–min fairness,* and *utility fairness.*

9.8.1 Maximizing Total Capacity

In designing a cellular system for mobile computing, maximum total capacity was traditionally used as the performance objective. In order to maximize the total capacity of a node, first the best channel conditions in a given slot should send the data. Nodes that are further away will less frequently satisfy this constraint, but will still have a very good and positive throughput due to the random part of fading. However, if a node is very far away from the base station, its average rate will be very small and essentially it will not be able to communicate.

A solution is found by assigning weights to each node rate such that a level of fairness is assured. The implicit assumption in this type of network is that an area with mobile nodes is well covered with base stations, so there is not a great variation in distances from the mobile nodes to the closest base stations. A similar direction was taken for high date rate (HDR) and code division multiple access (CDMA) networks. In the case of multihop wireless networks, the concept of assigning weights will not work because a node does not communicate with the closest base station but to an arbitrary destination in the network. A famous rate-based metric, based on the weighted sum of rates and used in multihop wireless settings, is a remedy for this type of problem.

9.8.2 Max–Min Fairness

Max–min fair rate allocation is defined to be the flow rate allocation in which every flow has a bottleneck link. This rate allocation can be obtained using the well-known water-filling algorithm. However, it is not always obvious how to generalize the notion of a bottleneck link and the water-filling approach to an arbitrary problem.

A simple example is a wireless ad hoc network. Even if every flow has a bottleneck link, the allocation may not be max–min fair. In particular, by decreasing the transmit power of some links, one can decrease the interference on the receiver of one of the bottleneck links and thus increase the link's rate. This way, a flow will loosen its bottleneck, and one might be able to increase its rate. It is difficult to define the concepts of a bottleneck link and of the water filling in the given example; furthermore, it is not obvious if the max–min fair rate allocation can be defined at all in this.

9.8.3 Utility Fairness

Utility fairness is a different approach to fairness in wired networking. Every user is assigned a utility that is a function of its rate. The goal of a network is to maximize the sum of utilities of all users to improve network performance. By selecting utility functions, a designer can achieve different trade-offs between efficiency and fairness. It can also be shown that TCP implements a form of utility fairness. Variants of utility fairness are used in existing wireless multihop network protocols.

9.9 Cross-Layer Protocols

SI NO.	CROSS-LAYER PROTOCOL	FOCUS OF THE PROTOCOL	REFERRED ALGORITHMS	CURRENT ALGORITHMS	SIMULATOR	LIMITATIONS OF PROTOCOL
1	A cross-layer optimization approach for efficient data gathering in wireless sensor network (WSN)	A novel cross-layer optimization approach that assumes a very simple MAC protocol and makes use of both routing MAC layers information to reduce congestion, improve delivery ratio, and optimize energy	Single-tree algorithm	Multiple-tree algorithm	Ns-2.2	Investigate different multiple tree construction algorithms
2	A cross-layer approach for energy-efficient MAC layer in WSN	Reduce the time spent on data moving (i.e., nodal processing time drops dramatically and thus energy consumption is reduced accordingly) Energy efficiency, scalability, adaptability to changes of network topology, collision avoidance, channel utilization, latency, throughput, and fairness	S-MAC D-MAC T-MAC	IEEE 802.11 DCF CL-MAC (cross-layer MAC protocol)	OPNET	This cross-layer design approach will be promising for MAC layer design. For further research, more thorough simulation, and theoretical analysis should conducted.

SI NO.	CROSS-LAYER PROTOCOL	FOCUS OF THE PROTOCOL	REFERRED ALGORITHMS	CURRENT ALGORITHMS	SIMULATOR	LIMITATIONS OF PROTOCOL
3	Cross-layer protocols for energy-efficient WSN	Focus on reducing the *energy loss* due to idle listening, control signaling congestion hot spots, packet collision, and *conserve the battery power*	Maximum throughput algorithm Maximum lifetime algorithm	Extended power-aware routing scheduling algorithm (*EPARS*)	*TinyOS*	Future sensors should be able to increase their operational lifetime substantially before requiring maintenance.
4	Cross-layer energy-efficient routing (XLE2R) for prolonged lifetime of WSN	Focus on *reducing power consumption in WSN* due to collision, overhearing control packet overhead, and ideal listening and over emitting	Dynamic source routing (DSR)	Cross-layer energy-efficient routing (*XLE2R*)	*OPNET*	Intend to implement this approach when the node is mobile and talk about packet loss.
5	Multicast lifetime maximization using network coding: a cross-layer approach	*Instead of using dynamic routing and changing paths during lifetime of the network, we investigate the scenarios in which the source communication with the destinations, using all feasible paths in static scheme. *Maximize the lifetime of energy-constrained wireless ad hoc for multicast application using network coding*	Maximum lifetime multicast problem (MLMP)	Maximum lifetime multicast problem with rate control (*MLMPRC*)		
6	A cross-layer approach for energy-efficient routing in mobile ad hoc networks	Power control link layer and routing protocol in network layer to *maximize the network lifetime*	E-AODV R-AODV	ER-AODV	NS-2.33	Mobility parameter and more layer interaction for routing protocol design
7	Cross-layer energy efficiency analysis and optimization in WSN	For successful delivery of all data generated by source nodes to the sink node with *minimal energy consumption using PPM and FSK*		PPM FSK	Numerical results	

SI NO.	CROSS-LAYER PROTOCOL	FOCUS OF THE PROTOCOL	REFERRED ALGORITHMS	CURRENT ALGORITHMS	SIMULATOR	LIMITATIONS OF PROTOCOL
8	An energy-efficient cross-layer clustering scheme for WSN	*Average the energy consumed by the nodes closer to the base station*, taking a cluster scheme that the cluster head nodes could be chosen based on the residual energy	AODV	Cluster scheme	NS-2	
9	Cross-layer design for QoS-aware energy-efficient data reporting in WSN	Adaptive routing metric that helps in *minimizing the energy consumption as well as meeting the real-time deadlines of the application*	Minimum cumulative energy routing	QoS energy-aware routing	NS-2	
10	An energy-efficient MAC protocol for cluster-based event-driven WSN application	The distributive feature of proposed MAC protocol is that it assigns a time slot to only one of the source nodes all with the same data sensed and to be sent. *Thus, it reduces data transmission redundancy and achieves energy savings.*	BMA TDMA E-TDMA	EA-TDMA	Analytical expression	
11	Review of the cross-layer design in wireless ad hoc and sensor network	In-depth *analysis of difficulties in cross-layer design and the latest cross-layer approach* for wireless ad hoc and sensor network	1. Difficulties in cross-layer design 2. Categorization of cross-layer approach 3. Layer trigger scheme with strict layering 4. Joint optimization schemes cooperated b/w multiple layers 5. Full cross-layer design sharing the overall network status			
12	Energy-efficient, reliable cross-layer optimization routing protocol for WSN	Feedback mechanism of communication control packets in MAC layer to address the issue of energy efficiency and reliability while designing a tree-based energy-efficient routing algorithm *to extend the network lifetime*	T-MAC	EERCP	NS-2	

SI NO.	CROSS-LAYER PROTOCOL	FOCUS OF THE PROTOCOL	REFERRED ALGORITHMS	CURRENT ALGORITHMS	SIMULATOR	LIMITATIONS OF PROTOCOL
13	Self-optimized autonomous routing protocol for WSNs with cross-layer architecture	Energy level and velocity metrics are trade-in from physical layer to network layer while *discovering an optimal route and also in initialization process*	ACO IAR ADR SC	BIOSARP	NS-2	The building and testing of a given self-organized protocol in the real WSN test bed

Problems

9.1 Define a cross layer.

9.2 Explain the cross-layer design principle along with general motivations for cross-layer design.

9.3 Describe the proposals involving cross-layer design for ad hoc networks.

9.4 Explain proposals for implementing cross-layer interactions.

9.5 Discuss the fundamental advantages offered by a layered architecture.

9.6 Explain some of the standard layered architectures with an example.

9.7 How is a performance objective met while designing a layered architecture?

9.8 Discuss the pitfalls of the cross-layer design approach.

9.9 Discuss some of the cross-layer protocols.

Bibliography

Boström, K. 2000. *Shannon's Source Coding Theorem*. 14469. Institute for Physics. University of Potsdam. Potsdam, Germany.

Canli, T., F. Nait-Abdesselam, and A. Khokhar. 2008. A cross-layer optimization approach for efficient data gathering in wireless sensor networks. *Networking and Communications Conference, INCC2008*. IEEE International, pp. 101–106.

DeCleene, B., V. Firoiu, M. Dorsch, and S. Zabele. 2005. Cross-layer protocols for energy-efficient wireless sensor networking. *Military Communications Conference, MILCOM 2005*. IEEE, 3:1477–1484.

De Couto, D. et al. 2003. A high-throughput path metric for multi-hop wireless routing. *Proceedings of MOBICOM*.

Dong, L., X. Wang, and S. Li. 2010. An energy efficient cross-layer clustering scheme for wireless sensor network. *Wireless Communications Networking and Mobile Computing (WiCOM), 2010 6th International Conference*, pp. 1–5.

Fang, Y., and A. B. McDonald. 2002. Cross-layer performance effects of path coupling in wireless ad hoc networks: Power and throughput implications of IEEE 802.11 MAC. *Proceedings of IEEE International Performance, Computing, and Communications Conference*, pp. 281–290.

Jarquin, O. S. 2004. Multi-hop radio networks in rough terrain: Some traffic-sensitive MAC algorithms. In *Energy-efficient link assessment in wireless sensor networks, Proceedings InfoCom, 2004*, ed. Keshavarzian et al.

Kawadia, V., and R. Kumar. 2005. A cautionary perspective on cross-layer design. *IEEE Wireless Communications* 12:3–11.

Khan, Z. A., M. Auguin, and C. Belleudy. 2010. Cross-layer design for QoS aware energy efficient data reporting in WSN. *Wireless Communication Systems (ISWCS), 2010 7th International Symposium*, pp. 526–530.

Li, J. et al. 2003. Performance evaluation of modified IEEE 802.11 MAC for multi-channel multihop ad hoc networks. *Journal of Interconnection Networks* 4 (30): 345–359.

Liang, Y. 2010. Energy-efficient, reliable cross-layer optimization routing protocol for wireless sensor network. *Intelligent Control and Information Processing (ICICIP), 2010 International Conference*, pp. 493–496.

Nilsson, A. 2004. Performance analysis of traffic load and node density in ad hoc networks. Proceedings European Wireless, 2004.

Raisinghani, W. T. et al. 2002. Improving TCP performance over mobile wireless environments using cross-layer feedback. *Proceedings IEEE International Conference on Personal Wireless Communications*, pp. 81–85.

Raman, B. et al. 2001. Arguments for cross-layer optimizations in Bluetooth scatternets. *Proceedings Symposium on Applications and the Internet*, pp. 176–184.

Ren, X., and J. Zhang. 2010. Review of the cross-layer design in wireless ad hoc and sensor networks. *Wireless Communications Networking and Mobile Computing (WiCOM), 2010 6th International Conference*, pp. 1–3.

Shakkottai, S. et al. 2003. Cross-layer design for wireless networks. *IEEE Communications* 41:74–80.

Srivastava, V., and M. Motani. 2005. Cross-layer design: A survey and the road ahead. *IEEE Communications Magazine* 43:112–119.

Tang, Q., C. Sun, W. Huan, and Liang,Y. 2010. Cross-layer energy efficiency analysis and optimization in WSN. *Networking, Sensing and Control (ICNSC), 2010 International Conference*, pp. 138–142.

Toumpis, S., and A. J. Goldsmith. 2003. Performance, optimization, and cross-layer design of media access protocols for wireless ad hoc networks. *Proceedings IEEE ICC, 2003*.

Yu, X. 2004. Improving TCP performance over mobile ad hoc networks by exploiting cross-layer information awareness. *Proceedings MobiCom, 2004*.

Yu, Y. 2008. A cross-layer approach for energy efficient MAC layer in wireless sensor networks. *Wireless Communications, Networking and Mobile Computing, 2008. WiCOM '08. 4th International Conference*, pp. 1–3.

Yuen, W. et al. 2002. A simple but effective cross-layer networking system for mobile ad hoc networks. *Proceedings IEEE PIMRC*, 2002.

Zuniga, M., and B. Krishnamachari. 2004. Analyzing the transitional region in low power wireless links. *Proceedings SECON*, 2004.

10

APPLICATIONS AND
RECENT DEVELOPMENTS

10.1 Introduction

Wireless ad hoc networking has achieved significant growth in the current decade. Due to the increased use of the Internet in our daily lives and the success of second-generation cellular systems, mobile wireless data communication have advanced both in terms of technology and usage/penetration. This has attracted the attention of the research communities toward truly ubiquitous computing and communication. As a complement to traditional large-scale communication, the demand for short-range data transactions is increasing fast as most man–machine communication as well as oral communication between human beings occurs at distances of less than 10 meters. This requires the exchange of huge volumes of data between the communicating parties.

Wireless communication can be deployed quickly and at highly reduced cost with the introduction of developing radio technologies (such as Bluetooth) that use license-exempted frequency bands. Many computing and communication devices, such as personal digital assistants (PDAs) and mobile phones, have gained immense popularity due to their decreasing price, portability, and usability in the context of an ad hoc network. With the continuous advancement of technology, these devices will become cheaper and highly feature rich.

Ad hoc wireless networks are self-configurable communication networks in which the nodes move frequently, and they do not rely on any fixed infrastructure for communication between them. The network topology changes frequently due to the highly dynamic nature of the nodes. Due to the limited range of transmission, the mobile devices in such networks act as the host as well as the router. When the destination is outside the direct transmission range of the source, the source node needs the help of the intermediate nodes for forwarding data to the destination. These networks have dynamic topology, bandwidth-constrained variable capacity wireless links, energy-constrained devices, and limited physical security. Due to the ad hoc nature of these networks, they are highly useful for commercial and military applications. Also, because of infrastructure-less operation, these networks are particularly useful for providing communication support where no communication infrastructure exists or the previous infrastructure has been totally destroyed and establishing a new infrastructure is not possible. They are especially useful in emergency situations, healthcare, home networking, and disaster recovery operations.

With the growing need of speedy and reliable access to the information, communication networks are playing a major role in our society. Due to the recent technological advancements in information technology and easy availability of cheap mobile communication devices, the use of wireless communication networks has increased rapidly. This has caused tremendous growth of wireless networks and development of new applications for various wireless network scenarios. The development of mobile ad hoc wireless networks has resulted from the huge demand of anytime/anywhere access to information. Wireless ad hoc networks are essentially communication networks.

These networks play a vital role in emergency scenarios such as disaster situations like earthquakes, flooding, etc., where the fixed network is destroyed and it is not possible to establish a new network quickly. As a result, rescue teams need to coordinate operations without the availability of fixed networks; mobile ad hoc networks are aptly suitable for such situations. Similarly, mobile ad hoc networks are highly essential for military operations where communication occurs in a hostile environment.

Due to the limited range of transmission of the mobile nodes, they need to act as the host as well as the router. In cases when the nodes are within their direct transmission range, they can communicate directly. But when the destination is outside the direct transmission range of the source, the source node has to receive help from the intermediate nodes in order to forward the packet to the destination. For providing such multihop routing capability to the nodes, some form of routing protocol, which can address a diverse range of issues such as low bandwidth, mobility, and low power consumption, is necessary in ad hoc networks. Due to the constant movement of the nodes, the mobile nodes in an ad hoc wireless network may move away from each other such that they are not able to maintain communication with others. In such a situation, an ad hoc network may be partitioned into two or more independent ad hoc networks. On the other hand, sometimes the mobile devices in two or more ad hoc networks may be in close proximity of each other and thus these networks may be combined into one larger ad hoc network. It becomes a huge challenge to manage such a highly dynamic networking environment.

In the field of mobile ad hoc networking, sensor networks have become hugely popular. The size of the mobile devices is very small in these sensor networks. These devices can be as small as the size of a grain of rice and are self-sufficient in all respects—transmitting, receiving, processing, and power. These sensors can be programmed according to the need of any given application.

This chapter discusses opportunities and challenges faced in development of ad hoc wireless networks. The next section discusses applications and opportunities that ad hoc wireless networks provide along with the challenges that they face.

10.2 Typical Applications

Commercial ad hoc networks are highly essential in situations where no infrastructure (fixed or cellular) is available. Examples include rescue operations in remote areas in cases of natural disasters, or when local coverage must be deployed quickly at a remote construction site. Ad hoc networking could also serve as wireless public access in urban areas, due to its quick deployment and self-organizing feature. At the local level, participants at a conference can form an ad hoc network that links their notebooks or palmtop computers for sharing information. Ad hoc networks can also be appropriate for applications in home networks, where devices can communicate directly to exchange information, such as audio/video, alarms, and configuration updates. These networks may also be useful for environmental monitoring, where the networks could be used to forecast water pollution or to provide early warning of an approaching tsunami.

A personal area network (PAN) can be constructed to simplify intercommunication between various mobile devices (such as a cellular phone and a PDA) and thereby eliminate the tedious need for cables. This can also extend the mobility provided by the fixed network (that is, mobile Internet protocol [IP]) to widen the coverage of ad hoc network domain.

10.2.1 PAN

A *personal area network* is a computer network established around an individual person. These networks typically involve a mobile computer, a cell phone, and/or a handheld computing device such as a PDA. A person who has access to a Bluetooth PAN can use the GPRS/UMTS mobile phone as a gateway to the Internet or to a corporate IP network, thus satisfying the need for anytime/anywhere access to the information while on the move. In addition, Bluetooth PANs can be interconnected with scatternets, thereby increasing the capacity. Figure 10.1 shows a scenario in which four Bluetooth PANs are used.

A PAN can encompass several different access technologies distributed among its member devices, which exploit the ad hoc functionality in the PAN. For instance, a notebook computer can have a wireless LAN (WLAN) interface (such as IEEE 802.11 or HiperLAN/2) that provides network access when the computer is used indoors. Thus, the PAN would benefit from the various access technologies residing in the member devices.

With the maturity of the PAN concept, new devices and new access technologies can be incorporated into the PAN framework. It

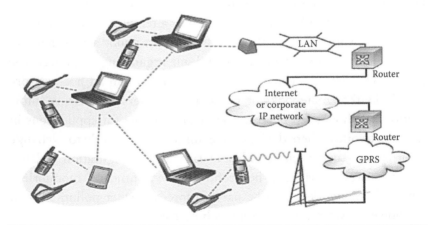

Figure 10.1   PAN scenario with four interconnected PANs, two of which have an Internet connection via a Bluetooth LAN access point and a GPRS/UMTS phone.

will also eliminate the need to create hybrid devices, such as a PDA–mobile phone combination, because the PAN network will instead implement wireless integration. In this way, the short-range communication networks such as Bluetooth will play a major role in introducing the flexibility represented by the PAN concept.

10.3 Applications and Opportunities

Ad hoc wireless networks are extremely useful for situations when there is a need for establishing a networking environment for a short time duration. These networks offer tremendous opportunities and can be used in numerous areas, particularly where a communication infrastructure is not available or it is difficult to establish the fixed infrastructure quickly. Typically, such applications include:

- Academic applications
- Defense (army, navy, air force) applications
- Industrial/corporate environment applications
- Healthcare applications
- Search and rescue operations in disaster situations
- Vehicular ad hoc networks

There are many other applications where ad hoc wireless networks can be utilized.

10.3.1 Academic Environment Applications

Due to their ability to deploy easily and quickly, mobile ad hoc networks have become extremely popular within the academic community. Most academic institutions have installed wireless communication networks in their campuses so that both students and teachers can avail themselves of the benefits of the ad hoc networking environment. Due to the huge popularity of laptops, PDAs, smart phones, etc. among students, they can easily get connected to an existing ad hoc network or form a new ad hoc network quickly.

Such an environment makes the interaction among the students and faculty very convenient. For instance, a teacher entering a class with his or her laptop can easily form an ad hoc wireless communication network with the students that have their own devices. As part of the same ad hoc network, the teacher can easily share lecture notes and assignments with the students and the students can submit their assignments and send their queries to the teacher. Sharing information among the class participants can be as easy as a click of a key on the keyboard. Due to the inherent mobile nature of the nodes in a mobile ad hoc network, these networks can also be of immense use while on a field trip and or industrial visit. Staying in touch cannot be any easier than this. Also, during conferences or workshops, the participants can form a temporary ad hoc network between them that will allow them to share their research materials, slides, etc.

10.3.2 Defense Applications

Defense operations take place in inhospitable terrains where communication infrastructure is not available. Wireless ad hoc and sensor networks are of urgent necessity in such situations. Different units involved in defense operations also need to maintain communication with each other while they are on the move. For example, air force planes flying in a formation may establish a temporary ad hoc wireless network for communicating with each other and for sharing images and data among themselves. In battlefield scenarios, army groups on the move can also use ad hoc wireless networks for communicating among themselves. A nice feature of such a communication environment is that the ad hoc network moves as the individuals move or the planes fly.

Information gathering is a highly crucial application of ad hoc wireless (particularly sensor) networks in military operations. Sensor networks can be deployed for intelligence gathering for defense purposes and they prove to be extremely effective. The sensors used for such applications are essentially disposable and are used for an application once. They can be deployed by air or by other appropriate means in large quantities over a selected area chosen for intelligence gathering. Because of their small size, these sensors will remain suspended in the air for some time. During that time, they can gather information according to their programmed logic, process the information, share among other nearby sensors, reach a consensus, and transmit information to a central server. The central processing facility can then analyze the gathered information and a decision about the next step can be made based on the analysis.

Due to the rapid advancements in semiconductor technologies, the size of the electronic devices is getting smaller. At the same time, these devices are equipped with higher and higher processing power on tiny chips. These advancements have led to the development of wearable computers. The idea of wearable computers is not that new, but the idea of a smart dress (that consists of many tiny computers [or sensors]) is relatively recent. In a smart dress, tiny computers are connected by tiny wires or by wireless means. They can exchange information with each other, process the gathered information, and take an appropriate action according to their program. A smart dress may be programmed to monitor certain conditions and vital signs of an individual on a regular basis. This could become very useful for defense personnel in combat situations. The monitored information can be processed and appropriate action can be taken by the dress, if needed. A smart dress may even be able to indicate the exact location of the problem and call for help if the seriousness of the situation warrants that.

10.3.3 Industrial Environment Applications

In industrial environments, wireless ad hoc networks play a major role in establishing communication between the different devices as well as between the employees for proper coordination of project activities. They are especially useful for manufacturing environments. The need for interconnection between various electronic devices requires the deployment of networking facilities between them. Connecting these large numbers of devices using wired connection leads to cluttering. Also, a huge amount of space is wasted. This not only creates safety hazards but also adversely affects reliability.

These problems can be eliminated with the use of wireless communication networks. The infrastructure-less mode of operation of the ad hoc networks helps to reduce the cost. Also, such networks can be established within a very short span of time. The support for mobility of the ad hoc mobile networks allows the devices to be relocated easily. Also, the dynamic nature of the networks allows easy reconfiguration of the networks based on the requirements. Employees carrying handheld computers, smart phones, etc. can easily and quickly form an ad hoc network among themselves that enables them to maintain proper communication and to work together without the need to get all the employees assembled in a meeting room. This ensures proper coordination among the different work units.

10.3.4 Healthcare Applications

In critical and emergency scenarios, exchanging information between the patient and the healthcare facilities is very helpful. A healthcare professional, in many situations, will be able to diagnose properly and accordingly be able to prepare a better treatment plan for an individual if he or she has access to video information rather than just audio or textual information. For example, with the help of video information, a doctor may better assess the reflexes and coordination capability of a patient.

In a similar way, a doctor can judge properly the severity of the injuries of a patient by visual information rather than by just audio or other descriptive information. Real-time ultrasound scans of a patient's kidneys, heart, or other organs may prove to be very helpful in preparing a proper treatment plan for a patient who is being transported to a hospital prior to his or her arrival there. Such information can be transmitted through wireless communication networks from an ambulance to other healthcare professionals who are currently scattered at different places but are converging toward the hospital for treating the patient being transported.

The concept of smart dress can also be used for monitoring health conditions of patients. Such dresses may become very useful for providing healthcare for our elderly population. Sometimes an ad hoc wireless network can be established within a (smart) home equipped with various sensors. This may be very useful for careful monitoring of homebound patients. On the basis of the information exchanged between various sensors monitoring the patient, some basic decisions may be taken that may be highly beneficial to the elderly population. These activities include recognizing falls of human beings, monitoring the movement patterns inside a home, and recognizing an unusual situation and informing a relevant agency in order to ensure appropriate and timely help, if needed.

10.3.5 Search and Rescue Applications

Ad hoc wireless networks can prove to be of enormous use for search and rescue operations in cases of natural disasters such as an earthquake, hurricane, flood, etc. In general, disasters leave a large population without power and communication capabilities due to the destruction of the infrastructure. Ad hoc wireless networks can be deployed in such situations quickly without the help of any fixed infrastructures and can provide communications among various relief organizations for coordinating their rescue activities. Wireless sensor networks can be of immense use in conducting searches for survivors and providing care in a timely manner.

Rescue operations also use robots in searching for survivors. Wireless ad hoc networks can be used to establish communication between these robots for coordinating their activities. Depending on the size of the area affected by a disaster, an appropriate number of robots can be deployed. An ad hoc network can be established between them for searching the area and for information gathering in the shortest possible time. The information thus collected can be analyzed and processed, and appropriate relief/help can accordingly be readily directed where essential.

10.3.6 Vehicular Ad Hoc Networks

With the huge proliferation of wireless technologies, mobile ad hoc networks have found widespread applications in the automobile industry. Nowadays, cars are equipped with different kinds of sensors, microcomputers, and wireless devices. This allows the formation of a new kind of mobile ad hoc network between the nearby moving vehicles or between vehicles and the roadside infrastructure. These networks are known as vehicular ad hoc networks.

These networks are self-organizing and multihop and enable the exchange of data between the users in nearby vehicles. With the help of these networks, intelligent transportation systems (ITSs) can be built that provide several benefits to users in terms of road safety, collision prevention, traffic scenario monitoring, congestion avoidance, infotainment, etc. Vehicular ad hoc networks are characterized by the high mobility of the vehicles, resulting in frequent and dynamic changes in the network topology. Due to this highly dynamic scenario, routing is considered a challenging task in vehicular ad hoc networks. Hence, the design of efficient routing protocols to suit the needs of these networks has become a key issue.

10.4 Challenges

Although mobile ad hoc networks offer significant advantages over wired networks, there are several challenges that need to be addressed properly for fully obtaining the benefits. The nodes in a mobile ad hoc network are constrained due to the following:

- Limited battery power and longevity
- Limited communication bandwidth and capacity
- Information security
- Size of the mobile devices
- Communication overhead
- Highly dynamic topology

Mobile communication devices are restricted in terms of the amount of power available. Due to their frequent mobility patterns, they cannot be connected to a constant source of power. These devices are powered by small batteries, which can supply only a limited amount of energy. As a result, the mobile devices need to minimize the power usage in order to increase their lifetime. This has attracted the attention of the research community and the industry to the design of devices that consume less power and can adjust the strength of communication signals based on the distance between communicating points. In addition, efficient signal processing techniques and algorithms are being developed that will reduce power usage significantly.

Due to the limited capacity of the communication medium, wireless networks suffer from bandwidth problems. This restricts the amount of information to be transmitted over a particular time duration. Efficient and innovative transmission techniques need to be invented to utilize the available bandwidth effectively and to increase the capacity.

The use of efficient transmission techniques, such as CDMA, and the structure of cellular communication mechanisms are very helpful in effectively using the available capacity. However, there is still the need of more research in this area to provide more efficient mechanisms for better utilization of the available communication bandwidth in the wireless communication environment.

Due to less security of the wireless communication medium, ad hoc wireless networks are more prone to security threats than their wired counterparts. Achieving the desired information security level requires additional processing overhead, which will be a major problem for the mobile nodes due to their limited processing power. It also requires additional bandwidth for secured transmission. A significant amount of research is already going on for discovering mechanisms for ensuring secure information transfer while at the same time not being prohibitive in terms of overhead.

With the advancements in semiconductor technologies, higher numbers of electronic components can be placed on smaller chips, which has led to the development of mobile devices that are more powerful and less power consuming.

Due to the limited capacity of the wireless medium, minimizing the communication overhead for information transfer in ad hoc wireless networks has become one of the biggest and most formidable challenges. In order to ensure proper delivery of information, a path needs to be established between the source and the destination. Moreover, a procedure that will ensure the sharing of a common pool of resources, such as bandwidth, has to be established.

One of the biggest challenges for designing routing protocols for mobile ad hoc networks is to handle the highly dynamic topology of these networks. For a source node to send information to a destination node, the source must be able to find the location of the destination node as well as other intermediate nodes. But due to their highly dynamic nature, the mobile devices change their locations frequently. As a result, a route that is established at the initial phase of the information transfer between two mobile devices may not be the same at the later phase of the information exchange. In order to adapt to this highly dynamic scenario, the routing protocols must be dynamic and adaptive in nature and the nodes must maintain up-to-date routing information all the time.

The routing protocols for mobile ad hoc networks are basically of two types: proactive and reactive. In the case of a proactive approach, the nodes need to maintain up-to-date routing information to all the nodes all the time. This requires the nodes to exchange the routing information between them periodically. The advantage of this approach is that the devices will always have routes available to other devices. Timely and periodic exchange of routing information will ensure the availability of fresh routes. Moreover, due to the immediate availability of the routes, no time will be wasted in setting up the path between the source and the destination devices. This ultimately reduces the delivery time of the information from the source to the destination device. The disadvantage of the proactive approach is the high overhead due to the periodic exchange of routing information between all the devices, even when all the routes may not be required.

In the case of a reactive approach, the routes to the destination are determined on an on-demand basis. The advantage of this approach is that the overhead incurred will be reduced as only the routes that are needed will be discovered. There is no need of periodic information exchange between the devices. However, this approach will suffer from more waiting time because routes will not be immediately available. The initial path setup takes a significant amount of time and, during this time, no packet can be sent to the destination due to the unavailability of routes. In order to combine the advantages of these approaches, many hybrid routing mechanisms have been introduced. However, the problem of finding the shortest routes with minimum overhead remains an open challenge.

10.4.1 Security

Security is a major area of concern for mobile ad hoc networks due to the limited physical security of the wireless medium. Although some work has been done on security for MANETs, the research on MANET security is still in its early stage. The existing security schemes in MANET are basically attack based. Certain frequent attacks on mobile ad hoc networks are identified first, and then security schemes are developed to prevent only these known attacks. But these networks operate in the real world. Such scenarios keep on changing and, as a result, attackers may develop new types of security attacks. Due to the changing pattern of these attacks, the existing security schemes fail to keep the security of the system intact. One feasible solution to this problem would be to develop a multifence security solution. This type of solution offers protection against a broader area of malicious activities. When embedded into possibly every component in the network, it offers in-depth protection in the form of multiple lines of defense against many known and unknown security threats.

This kind of new approach to designing the security system is known as resiliency-oriented security design. It consists of several features:

- This kind of security system tries to cover a broader problem area. Due to a multiline defense architecture, it is capable of handling not only the malicious attacks of known patterns, but also network faults that occur due to node misconfiguration, extreme network overload, or operational failures. After careful observation, one can notice that all such faults, whether incurred by attacks by malicious users or by misconfigurations, share some common symptoms from both network and end-user perspectives. Therefore, on the basis of this common signature, the security system should be able to detect such attacks and take proper measures to thwart such attempts.

- In terms of solution space, we can see that cryptography-based techniques offer a subset of tool kits in the case of a resiliency-oriented design. We need to use other noncryptographically based techniques for ensuring resiliency. For example, more "protocol invariant" information may be piggybacked in the protocol messages. This allows the nodes participating in the message exchange to verify such information. Routing messages can be propagated through multiple paths by exploiting the rich connectivity of the network topology, and redundant copies of such messages may be checked to detect inconsistency of the operations of the protocol.

- In the case of resiliency-oriented design, the focus has shifted from conventional intrusion prevention to intrusion tolerance. In the case of MANETs, certain degrees of intrusions or malicious attacks will always occur in the real world. Therefore, the systems must be designed to be robust to protect against such security threats. Systems security should be designed with multiple levels of security so that the collapse of an individual fence will not cause the breakdown of the entire system. Even if an attacker is able to break through a particular security level, the entire system should continue to function, possibly with graceful degradation.

- Sometimes unexpected faults may occur and the solution should be able to handle such faults to some extent. We can ensure this by strengthening the correct operation mode of the network by implementing more redundancy at the protocol and system levels. At each step of the protocol operation, there must be checks to ensure that everything has been done correctly along the right track. Whenever we notice any deviation from valid operations, it should be treated with caution and an alarm should be raised. The system should query the identified source for further verification. This way, the protocol can distinguish right from wrong as it has complete knowledge about what is right but not necessarily knowing what is exactly wrong. This way the design can strengthen the correct operations and may even handle previously unknown threats tha may occur in runtime operations.

Researchers working in different fields, such as wireless networking, mobile systems, and cryptography, need to work collaboratively in order to develop an effective evaluation methodology and tool kits to ensure the security of wireless ad hoc networks.

10.5 Highlights of the Most Recent Developments in the Field

With recent advances in the electronics and telecommunication industries, electronic components have become cheaper, faster, and more reliable. This has caused tremendous growth in computing and communication technologies. Due to the huge availability and portability of highly mobile devices such as laptops, smart phones, PDAs, etc., there has been a huge demand for access to information while on the move.

Both coverage and wireless sensor networks are intrinsically multidisciplinary research topics. Therefore, a wide body of scientific and technological work is related to research presented in this chapter. In this section, we briefly cover only the most directly related areas: sensors, wireless ad hoc sensor networks, the coverage problem, and related sensor network problems such as location discovery and deployment.

10.5.1 Sensors

A sensor is a device that can sense and measure a change in the physical condition of the environment, such as change of air pressure, temperature, etc. Although we have been using sensors in various applications for a long time, sensors have recently found an even wider range of applications with the emergence of microelectromechanical system (MEMS) sensors, which offer small size, reduced cost, and high reliability, and due to the interconnection between the sensors and computer networks. Today, we find extensive use of sensors everywhere—from home applications to space flights.

10.5.2 Wireless Ad Hoc Sensor Networks

Due to their easy availability and reduced cost, wireless sensor networks have found a wide array of commercially viable applications. Because of their huge potential to be applied in a number of future applications, they have drawn the attention of the research community. The use of sensors in mobile ad hoc networks has practically caused a revolution in opening up a huge possibility in diverse fields of applications. Due to the infrastructure-less and self-configurable nature of wireless ad hoc networks, they can be deployed very quickly without the help of any fixed infrastructure and they show high adaptability in highly dynamic scenarios. Due to the integration of easily available, low-cost, power-efficient, and reliable sensors in nodes of wireless ad hoc networks equipped with significant computational and communication resources, a diverse range of research and engineering vistas has opened up. With this emerging area of applications of the sensors in our various facets of life come the various challenges related to the new technical problems, including the need for new operating systems, DSP algorithms, integration with biological systems, and low-power architectural designs.

10.6 Summary

With the huge influx of highly mobile and portable devices and due to the infrastructure-less mode of operation of mobile ad hoc networks, these networks are finding increasing applications in many areas, including disaster recovery, healthcare, defense, academic, and industrial environments. However, with the many advantages of the mobile ad hoc networks come many challenges that are still unresolved. The challenges are related to the constrained environment of the mobile ad hoc networks, which includes development of mechanisms for efficient use of limited bandwidth and channel capacity, developing smaller sized but feature-rich mobile devices, techniques for minimizing power consumption and hence extending network lifetime, developing algorithms for enhancing information security, and developing efficient routing procedures for finding better routes with less overhead. A significant amount of research is already going on in the field of mobile ad hoc networking and still there is a huge scope of research in order to meet the challenges for solving open problems.

Bibliography

Yang, H., H. Y. Luo, F. Ye, S. W. Lu, and L. Zhang. 2004. Security in mobile ad hoc networks: Challenges and solutions. UC Los Angeles. Retrieved from http://escholarship.org/uc/item/5p89k583

Index

Printed and bound by CPI Group (UK) Ltd, Croydon, CR0 4YY

21/10/2024

01777105-0008